Institute of Mathematical Statistics

LECTURE NOTES-MONOGRAPH SERIES
Shanti S. Gupta, Series Editor

Volume 9

Fu atistical Ex s
with ecision Theory

Lawr

Corne

Institute of Mathematical Statistics
Hayward, California

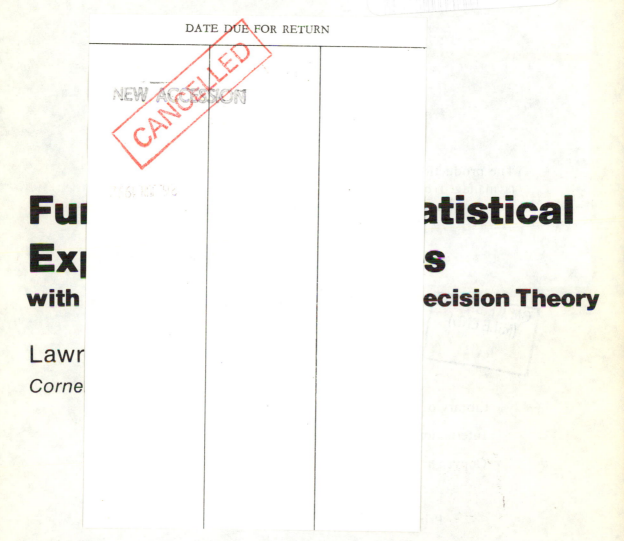

Institute of Mathematical Statistics

Lecture Notes–Monograph Series

Series Editor, Shanti S. Gupta, Purdue University

The production of the IMS Lecture Notes–Monograph Series is managed by the IMS Business Office: Nicholas P. Jewell, Treasurer, and Jose L. Gonzalez, Business Manager.

Library of Congress Catalog Card Number: 87–80020

International Standard Book Number 0–940600–10–2

Printed in the United States of America

To my family

for their love and understanding

PREFACE

I first met exponential families as a beginning graduate student. The previous summer I had written a short research report under the direction of Richard Bellman at the RAND Corporation. That report was about a dynamic programming problem concerning sequential observation of binomial variables. Jack Kiefer read that report. He conjectured that the properties of the binomial distribution used there were properties shared by all "Koopman-Darmois" distributions. (This is a name sometimes used for exponential families, in honor of the authors of two of the pioneering papers on the topic. See Koopman (1936), and Darmois (1935), and also Pitman (1936).)

Jack suggested that I recast the paper into the Koopman-Darmois setting. That suggestion had two objectives. One was the hope that viewing the problem from this general perspective would lead to a clearer understanding of its structure and perhaps a simpler and better proof. The other objective was the hope of generalizing the result from the binomial to other classes of distributions, for example the Poisson and the gamma. (The resulting manuscript appeared as Brown (1965).)

These two objectives of clearer understanding and of possible generalization in statistical applications are the motivation for this monograph. Many if not most of the successful mathematical formulations of statistical questions involve specific exponential families of distributions such as the normal, the exponential and gamma, the beta, the binomial and the multinomial, the geometric and the negative binomial, and the Poisson among others. It is often informative and advantageous to view these mathematical formulations from the perspective of general exponential families.

These notes provide a systematic treatment of the analytic and probabilistic properties of exponential families. This treatment is constructed with a variety of statistical applications in mind. This basic theory appears in Chapters 1-3, 5, 6 and the first part of Chapter 7 (through Section 7.11). Chapter 4, the latter part of Chapter 7, and many of the examples and exercises elsewhere in the text develop selected statistical applications of the basic theory.

Almost all the specific statistical applications presented here are within the area of statistical decision theory. However, as suggested above the scope of application of exponential families is much wider yet. They are, for further example, a valuable tool in asymptotic statistical theory. The presentation of the basic theory here was designed to be also suitable for applications in this area. Exercises 2.19.1, 5.15.1-5.15.4 and 7.5.1-7.5.5 provide further background for some of these applications. Efron (1975) gives an elegant example of what can be done in this area.

Some earlier treatments of the general topic have proved helpful to me and have influenced my presentation, both consciously and unconsciously. The most important of these is Barndorff-Nielsen (1978). The latter half of that book treats many of the same topics as the current monograph, although they are arranged differently and presented from a different point-of-view. Lehmann (1959) contains an early definitive treatment of some fundamental results such as Theorems 1.13, 2.2, 2.7 and 2.12. Rockafellar (1970) treats in great detail the duality theory which appears in Chapters 5 and 6. I found Johansen (1979) also to be useful, particularly in the preparation of Chapter 1.

The first version of this monograph was prepared during a year's leave at the Technion, Haifa, and the second was prepared during a temporary appointment at the Hebrew University, Jerusalem. I wish to express my gratitude to both those institutions and especially to my colleagues in both departments for their hospitality, interest, and encouragement. I also want to acknowledge the support from the National Science Foundation which I received throughout the preparation of this manuscript.

I am grateful to all the colleagues and students who have heard me lecture on the contents or have read versions of this monograph. Nearly all have made measurable, positive contributions. Among these I want to specially thank Richard Ellis, Jiunn Hwang, Iain Johnstone, John Marden, and Yossi Rinott who have particularly influenced specific portions of the text, Jim Berger who made numerous valuable suggestions, and above all Roger Farrell who carefully read and critically and constructively commented on the entire manuscript. The draft version of the index was prepared by Fu-Hsieng Hsieh.

Finally, I want to thank the editor of this series, Shanti Gupta, for his gentle but persistent encouragement which made an important contribution to the completion of this monograph.

TABLE OF CONTENTS

Chapter 1. Basic Properties

STANDARD EXPONENTIAL FAMILIES

1.1 Definitions (Standard Exponential Family): Let ν be a σ-finite measure on the Borel subsets of R^k. Let

(1) $$N = N_\nu = \{\theta : \int e^{\theta \cdot x} \nu(dx) < \infty\} .$$

Let

(2) $$\lambda(\theta) = \int e^{\theta \cdot x} \nu(dx) .$$

(Define $\lambda(\theta) = \infty$ if the integral in (2) is infinite.) Let

$$\psi(\theta) = \log \lambda(\theta) ,$$

and define

(3) $$p_\theta(x) = \exp(\theta \cdot x - \psi(\theta)) , \quad \theta \in N .$$

Let $\Theta \subseteq N$. The family of probability densities

$$\{p_\theta : \theta \in \Theta\}$$

is called a k-dimensional *standard exponential family* (of probability densities). The associated distributions

$$P_\theta(A) = \int_A p_\theta(x) \nu(dx) , \quad \theta \in \Theta$$

are also referred to as a standard exponential family (of probability distributions).

N is called the *natural parameter space*. ψ has many names. We will call it the *log Laplace transform* (of ν) or the *cumulant generating function*. $\theta \in \Theta$ is sometimes referred to as a *canonical parameter*, and

$x \in X$ is sometimes called a *canonical observation*, or value of a *canonical statistic*.

The family is called *full* if $\Theta = N$. It is called *regular* if N is open, i.e. if

$$N = N^\circ$$

where N° denotes the interior of N, defined as int. $N = \{\cup Q: Q \subset N, \ Q \text{ is open}\}$.

As customary, let the support of ν (supp ν) denote the minimal closed set $S \subset R^k$ for which $\nu(S^{\text{comp}}) = 0$. Let

$$H = \text{convex hull (supp } \nu) = \text{conhull (supp } \nu) . \tag{4}$$

and let $K = K_\nu = \bar{H}$. K is called the *convex support* of ν. (The convex hull of a set $S \in R^k$ is the set $\{y: \exists \{x_i\} \subset S, \{\alpha_i\}, 0 < \alpha_i, \Sigma\alpha_i = 1 \ni y = \Sigma\alpha_i x_i\}$.)

For $S \subset R^k$ the dimension of S, dim S, is the dimension of the linear space spanned by the set of vectors $\{(x_1 - x_2): x_1, x_2 \in S\}$. A k-dimensional standard family is called *minimal* if

$$\dim N = \dim K = k . \tag{5}$$

Note that if K is compact then $N = R^k$, so that the family is regular.

(The exponential families described above can be called finite dimensional exponential families. Various writers have recently begun to investigate infinite dimensional generalizations. See Soler (1977), Mandelbaum (1983), and Lauritzen (1984) for some results and references.)

Standard exponential families abound in statistical applications. Often a reduction by sufficiency and reparametrization is, however, needed in order to recognize the standard exponential family hidden in specific settings. Here are two of the most fruitful examples.

<u>1.2 Example (Normal samples):</u> Let $Y_1, \ldots, Y_n$ be independent identically distributed normal variables with mean μ and variance σ^2. Thus, each Y_i has

density (relative to Lebesgue measure)

(1) $$\phi_{\mu,\sigma^2}(y) = (2\pi\sigma^2)^{-\frac{1}{2}} \exp(-(y-\mu)^2/2\sigma^2)$$

and cumulative distribution function Φ_{μ,σ^2} . Consider the statistics

$$\bar{Y} = n^{-1} \sum_{i=1}^{n} Y_i$$

$$S^2 = n^{-1} \sum_{i=1}^{n} (Y_i - \bar{Y})^2$$

$$X_1 = \bar{Y} , \quad X_2 = n^{-1} \sum_{i=1}^{n} Y_i^2 = S^2 + \bar{Y}^2 .$$

The joint density of $Y = Y_1,\ldots,Y_n$ can be written in two distinct revealing ways, as

(2) $$f_{\mu,\sigma^2}(y) = (2\pi\sigma^2)^{-n/2}\exp(-ns^2/2\sigma^2 - n(\bar{y}-\mu)^2/2\sigma^2) ,$$

or as

(3) $$f_{\mu,\sigma^2}(y) = (2\pi\sigma^2)^{-n/2}\exp((n\mu/\sigma^2)x_1 + (-n/2\sigma^2)x_2)\exp(-n\mu^2/2\sigma^2) .$$

From the first of these one sees that $\bar{Y}$ and S^2 are sufficient statistics. (One can also derive from this expression that $\bar{Y}$ and S^2 are independent (see sections 2.14 - 2.15) with $\bar{Y}$ being normal mean μ, variance σ^2/n and $V = S^2$ being $(\sigma^2/n)\cdot\chi^2_{n-1}$ -- i.e. having density

(4) $$f(v) = (n/2\sigma^2)^{m/2}(\Gamma(m/2))^{-1} v^{(m/2 -1)} \exp(-nv/2\sigma^2)\chi_{(0,\infty)}(v)$$

with $m = n-1$.)

$X = (X_1,X_2)$ is also sufficient. This can be seen from the factorization (3), or from the fact that X is a 1-1 function of $(\bar{Y},S^2)$. Let ν denote the marginal measure on R^2 corresponding to X -- i.e.

$$\nu(A) = \int_{(x_1,x_2)\in A} dy_1 \ldots dy_n .$$

(It can be checked that when $n \geq 2$, $\nu(dx) = (\frac{n}{2})^{n/2} (\pi^{\frac{1}{2}}\Gamma((n-1)/2))^{-1}(x_2-x_1^2)^{\frac{n-3}{2}} dx$ over the region $K = \{(x_1,x_2): x_1^2 \leq x_2\}$. When $n = 1$ ν is supported on the curve $\{(x_1,x_2): x_1^2 = x_2\}$.) Then the density of X relative to ν is

(5) $$p_{\theta_1,\theta_2}(x) = \exp(\theta_1 x_1 + \theta_2 x_2 - \psi(\theta))$$

with

$$\theta_1 = n\mu/\sigma^2 , \qquad \theta_2 = -n/2\sigma^2$$

and

$$\psi(\theta) = -\theta_1^2/4\theta_2 - (n/2)\log(-2\theta_2/n) .$$

Thus the distributions of the sufficient statistic form a 2 dimensional exponential family with canonical parameters (θ_1,θ_2) related to the original parameters as above.

This family is minimal. The natural parameter space is

$$N = \{(\theta_1, \theta_2) : \theta_1 \in R , \quad \theta_2 < 0\} .$$

The above can of course be generalized to multivariate normal distributions. See Example 1.14.

<u>1.3 Example</u> (Multinomial distribution):

Let $X = (X_1,\ldots,X_k)$ be multinomial (N,π) -- that is

$$\Pr\{X = x\} = \binom{N}{x_1,\ldots,x_k}\Pi\, \pi_i^{x_i} .$$

Let ν be the measure concentrated on the set $\{x : x_i$ integers, $x_i \geq 0$, $i=1,\ldots,k$, $\sum_{i=1}^{k} x_i = N\}$, and given by

(1) $$\nu(\{x\}) = \binom{N}{x_1,\ldots,x_k} = \frac{N!}{x_1!\ldots x_k!} .$$

Then the density of X relative to ν is

(2) $$p_\theta(x) = \exp(\sum_{i=1}^{k} \theta_i x_i - \psi(\theta))$$

where

(3) $$\theta_i = \log \pi_i \qquad i=1,\ldots,k$$

and

(4) $$\psi(\theta) = N \log\left(\sum_{i=1}^{k} e^{\theta_i}\right) .$$

This is a k dimensional exponential family with canonical statistic X . Its canonical parameter is related to π by (3). It has parameter space

(5) $$\Theta = \{(\log \pi_i) : 0 < \pi_i, \Sigma\pi_i = 1\} .$$

Note that this exponential family is not full. The full family has densities $\{p_\theta\}$ as above with $\Theta = N = R^k$. (For Θ as in (5) $\psi(\theta) \equiv 0$, however ψ as defined in (4), rather than $\psi \equiv 0$, is the appropriate cumulant generating function, as defined in 1.1(3) for the full family.) However

(6) $$p_\theta = p_{\theta + a1}$$

for all $a \in R$ where $1' = (1,\ldots,1)$. Hence expanding this family to be a full family does not introduce any new distributions.

The above phenomenon is related to the fact that the above family is not minimal since $\dim K = k-1 < k$. To reduce to a minimal family let $X^* \in R^{k-1}$ be given by $(X_1,\ldots,X_{k-1})$. Then X^* is sufficient. (In fact, it is essentially equivalent to X since $X_k = N - \sum_{i=1}^{k-1} X_i^*$ a.e.(ν) .) Let $\theta^* \in R^{k-1}$ be given by $\theta_i^* = \theta_i - \theta_k$, and let $\nu^*(\{x^*\}) = \binom{N}{x_1^*,\ldots,x_{k-1}^*, N-\sum_{i=1}^{k-1} x_i^*}$. Then the density of X^* relative to ν^* is

(7) $$p_{\theta^*}^*(x^*) = \exp\left(\sum_{i=1}^{k-1} \theta_i^* x_i^* - \psi^*(\theta^*)\right)$$

where

(8) $$\psi^*(\theta^*) = N \log\left(1 + \sum_{i=1}^{k-1} e^{\theta_i^*}\right) .$$

This is a full minimal standard exponential family with $N = R^{k-1}$.

Note that

$$\pi_i = \exp(\theta_i^*)/(1 + \Sigma\exp(\theta_i^*)) \qquad i=1,\ldots,k-1 ,$$

(9)

$$\pi_k = 1/(1 + \Sigma\exp(\theta_i^*)) .$$

Here, each different $\theta^* \in R^{k-1} = N$ corresponds to a different distribution.

Reductions by reparametrization and sufficiency like those in the above examples are frequent in statistical applications. Together with proper choice of the dominating measure, ν, they lead to the representation of problems involving exponential families in terms of problems involving standard exponential families. This is formally explained in the next few paragraphs.

1.4 Definition:

Let $\{F_\omega : \omega \in \Omega\}$ be a family of distributions on a probability space $\mathcal{Y},\mathcal{B}$. Suppose $F_\omega \ll \mu$, $\omega \in \Omega$. Suppose there exist functions

$$C : \Omega \to (0,\infty)$$

$$R : \Omega \to R^k$$

$$T : \mathcal{Y} \to R^k \quad \text{(Borel measurable)}$$

$$h : \mathcal{Y} \to [0,\infty) \quad \text{(Borel measurable)}$$

such that

(1) $$f_\omega(y) = \frac{dF_\omega}{d\mu} = C(\omega)h(y)\exp(R(\omega) \cdot T(y)) .$$

Then $\{F_\omega\}$ (or, $\{f_\omega\}$) is called a *k dimensional exponential family* of distributions (or, of densities).

1.5 Proposition:

Any k dimensional exponential family (1.4(1)) can be reduced by sufficiency, reparametrization, and proper choice of ν to a k dimensional standard exponential family (1.1(3)). The sufficient statistic is $X = T(Y)$, and its distributions form an exponential family with canonical parameter $\theta = R(\omega)$.

Proof: $X = T(Y)$ is sufficient by virtue of 1.4(1) and the Neyman factorization theorem. (See e.g. Lehmann (1959) Chapter 2 Theorem 8.) Let $\mu^*(dy) = h(y)dy$ and let $\nu(A) = \mu^*(T^{-1}(A))$ for Borel measurable sets $A \subset R^k$. Then the induced densities of X with respect to ν exist and have the desired form 1.1(3) with $\theta = R(\omega)$ and $\psi(\theta) = -\log C(R^{-1}(\theta))$. (Note that if $R(\omega_1) = R(\omega_2)$, then $f_{\omega_1} = f_{\omega_2}$ and hence $C(\omega_1) = C(\omega_2)$.) ||

In spite of appearances the above reduction process is not really unique. Any standard exponential family can be transformed to a different, but equivalent, form by linearly transforming X and Θ with linked nonsingular affine transformations. This is described in the following proposition.

1.6 Proposition:

Let $\{p_\theta\}$ be a k-dimensional standard exponential family. Let M be a non-singular $k \times k$ matrix and let

$$Z = MX + z_0$$

(1)

$$\phi = (M')^{-1}\theta + \phi_0 .$$

Then the distributions of Z also form a k-dimensional standard exponential family which is equivalent to the original family.

Proof: The equivalency assertion is immediate since the transformations (1) are 1-1. Furthermore, the density of Z relative to the measure ν_2 defined by $\nu_2(A) = \nu(M^{-1}(A - z_0))$ is

$$\exp(\theta' x(z) - \psi(\theta))$$

(2) $$= \exp((\phi - \phi_0)' MM^{-1}(z - z_0) - \psi(M'(\phi - \phi_0)))$$

$$= \exp(\phi' z - \psi(M'(\phi - \phi_0)) + \phi' z_0 - \phi_0' z + \phi_0 \cdot z_0).$$

(By definition $A - z_0 = \{x : \exists\, z \in A,\ \ x = z - z_0\}$.)

Let $\nu_1(dz) = \exp(-\phi_0' z)\nu_2(dz)$ and $\psi_1(\phi) = \psi(M'(\phi - \phi_0)) - \phi' z_0 + \phi_0 \cdot z_0$.

The densities of Z relative to ν_1 are

(3) $$\exp\{\phi' z - \psi_1(\phi)\} ,$$

which, as claimed, form a k parameter exponential family. The natural parameter space for this family is $M'^{-1}\Theta + \phi_0$ and the cumulant generating function is ψ_1 . ||

Proposition 1.6 shows that one may apply an arbitrary affine transformation either to Θ or to X. In this way one may assume without loss of generality that Θ (or X) lies in a convenient position in R^k . One application of this process will be discussed at some length in Section 3.11, and such transformations will be used wherever convenient.

MARGINAL DISTRIBUTIONS

The proof of Proposition 1.6 yields a statement about marginal distributions generated under linear projections by standard exponential families. The result is important in its own right, and useful in the proof of Theorem 1.8, as well.

Some preliminary remarks will be helpful. Let $\mathcal{M}_1 : R^k \underset{\text{onto}}{\rightarrow} R^m$ be a linear map. $\mathcal{M}_1$ is represented by an $(m\times k)$ matrix, M_1, of rank m. There is then a linear map $\mathcal{M}_2 : R^k \underset{\text{onto}}{\rightarrow} R^{k-m}$ which is orthogonally complementary to $\mathcal{M}_1$ -- that is, the rows of the corresponding $((k-m)\times k)$ matrix, M_2, of rank $(k-m)$ are orthogonal to those of M_1 . (The rows of M_2 can be chosen to be orthonormal, but that is not necessary here.) Let M denote the $(k\times k)$ nonsingular matrix $M = \binom{M_1}{M_2}$. If $x \in R^k$ then $Z = Mx$ can be written as $\binom{Z_1}{Z_2}$ with $Z_1 \in R^m$, $Z_2 \in R^{m-k}$.

Let $M = \binom{M_1}{M_2}$ as defined above. Then M^{-1} exists and can be written as

(1) $$M^{-1} = (M_1^-, M_2^-)$$

where M_1^- is $(k\times m)$, M_2^- is $(k\times(k-m))$ and

$$(2) \qquad (M_1^-)' M_2^- = 0$$

since M_1 and M_2 are orthogonally complementary.

Let $\theta \in R^k$ and $\phi = (M^{-1})'\theta = \binom{(M_1^-)'}{(M_2^-)'}\theta = \binom{\phi_1}{\phi_2}$. Then

$$(3) \qquad \theta' x = \theta' M^{-1} M x$$

$$= \phi_1' z_1 + \phi_2' z_2$$

by (2). For typographical reasons let $M_i'^- = (M_i^-)'$.

The special case where $M_1(x_1,\ldots,x_k) = (x_1,\ldots,x_m)$ is worth noting. Here

$$(4) \qquad M_1 = (I_{m\times m}, \; 0_{m\times(k-m)}) = M_1'^-$$

$$M_2 = (0_{(k-m)\times m}, \; I_{(k-m)\times(k-m)}) = M_2'^-$$

Somewhat more generally, if the rows of M_1 and M_2 are orthonormal then

$$(4') \qquad M_1 = M_1'^- \qquad M_2 = M_2'^-$$

1.7 Theorem:

Consider a standard exponential family. Let $M_1 : R^k \underset{\text{onto}}{\rightarrow} R^m$ and $\theta = M'\binom{\phi_1}{\phi_2}$ as described above. Fix $\phi_2^0 \in M_2'^-(N) \subset R^{k-m}$. Consider the family of distributions of $Z_1 = M_1 X$ over the parameter space $\Phi_{\phi_2^0} = M_1'^-(\{\theta \in \Theta : M_2'^-\theta = \phi_2^0\})$. These form an m dimensional standard exponential family generated by the marginal measure defined by

$$(5) \qquad \nu_{\phi_2^0}(A) = \int_{M_1^{-1}(A)} \exp(\phi_2^{0\prime} M_2 x)\nu(dx) .$$

The natural parameter space for this family is $N_{\phi_2^0} = M_1'^-(\{\theta \in N : M_2'^-\theta = \phi_2^0\})$.

The statistic Z_1 is sufficient for the family of densities $\{p_\theta(x) : M_2'^-\theta = \phi_2^0\}$.

Proof: A direct proof is as easy as an appeal to Proposition 1.6. The density of Z relative to the appropriate dominating measure $\nu(M^{-1}\,\cdot)$ is

(6) $$\exp(\theta'x - \psi(\theta)) = \exp(\phi_1 \cdot z_1 + \phi_2 \cdot z_2 - \psi(M'\phi)) .$$

When $\phi_2 = \phi_2^0$ the factor $\exp(\phi_2^0 \cdot z_2)$ can be absorbed into the dominating measure, yielding $\nu_{\phi_2^0}(\cdot)$ as defined in (5). The resulting family of densities is the standard exponential family claimed in the statement of the theorem. (Note that (6) also provides a formula for the cumulant generating function of this family.)

The assertions concerning $N_{\phi_2^0}$ and sufficiency follow from (6), with $\phi_2 = \phi_2^0$, and the Neyman factorization theorem. ||

For the special case where M_1 is as described in (4), one sees that for fixed $\theta_{k+1},\ldots,\theta_m$ the distributions of $Z_1 = (X_1,\ldots,X_k)$ form an exponential family.

Note that the theorem does not say that the family of distributions of $Z_1 = M_1X$ form a standard exponential family with natural parameter ϕ_1 if the parameter θ ranges over *all* of Θ . In fact such a claim is generally false unless Θ is of dimension $\leq m$ and satisfies

(7) $$\Theta \subset \{\theta : M_2'^{-}\theta = \phi_2^0\} \quad \text{for some } \phi_2^0 \in R^{k-m} ,$$

as will be the case in Theorem 1.9; or

(8) $$Z_1 \text{ and } Z_2 \text{ are independent for some } \theta \in \Theta .$$

(It will be seen in the next chapter that (8) implies independence of Z_1 and Z_2 for all $\theta \in \Theta$.)

(8) Remark. The preceding theorem may be given an alternative interpretation. Let L be a linear variety in R^k -- that is $L = x_0 + V$ for some m dimensional linear subspace $V \subset R^k$. Let $P : R^k \to L$ be any affine projection onto L -- that is, P is affine, $P^2 = P$, and P is the identity on L. Let Q denote the orthogonal projection onto $V^{\perp} = \{w \in R^k : v'w = 0 \;\forall\; v \in V\}$. Let

$\theta_{(2)} \in V^{\perp}$. Then the family of distributions of P(X) as θ ranges over $\{\theta \in N : Q\theta = \theta_{(2)}\}$ forms an exponential family.

To verify the above, note that there are linear isometries

$$S_1 : R^m \underset{\text{onto}}{\overset{\rightarrow}{\text{1-1}}} L \qquad S_2 : R^{k-m} \underset{\text{onto}}{\overset{\rightarrow}{\text{1-1}}} V^{\perp} .$$

The theorem applies to the maps $M_1 = S_1^{-1} \circ P$, $M_2 = S_2^{-1} \circ Q$, and yields a statement concerning the distributions of $M_1(X)$. This converts directly to the above statement about the distributions of $P(X) = S_1(M_1(X))$ over the appropriate parameter space since S_1 is a linear isometry, and S_1 and S_2 are orthogonal, etc.

1.8 EXAMPLE (Log-linear models): Consider a multinomial (N, π) variable as described in Example 1.3. Consider the family of distributions for which the natural parameter 1.3(3) satisfies

(1) $$\theta = B\beta + \theta_0 , \qquad \beta \in R^m$$

where B is a specified k×m matrix of rank m. Assume, in addition, that

(2) $$B = (1_k, B_{(2)})$$

where $1_k' = (1,\dots,1)$ and $B_{(2)}$ is $k \times (m-1)$ of rank (m-1). This is a *log-linear multinomial model*. The name derives from the fact that the linear constraint (1) can also be written in the form $\log \pi = B\beta$ where $(\log \pi)_i = \log \pi_i$, $i=1,\dots,k$. Condition (2) is imposed because $P_\theta = P_{\theta+a1}$, as noted in 1.3(6). Because of (2) for every $\beta_{(2)}' = (\beta_2,\dots,\beta_m)$ there is a unique $\beta_1 = \beta_1(\beta_{(2)})$ such that

(3) $$\sum_{i=1}^{k} \pi_i = \sum_{i=1}^{k} e^{\theta_i} = 1 .$$

Let $M_1 = B'$ and let $M = \binom{M_1}{M_2}$ as in 1.7. Theorem 1.7 yields that $Z_{(1)} = M_1X = B'X$ is a sufficient statistic. The distributions of $Z_{(1)}$ form an m-dimensional exponential family with corresponding natural parameter

$M_1^{\prime -}\theta = \beta + B^{-}\theta_0$. This family is not minimal since $(Z_{(1)})_1 = N$ w.p. 1 . As in Example 1.3 one may reduce to an equivalent minimal family having dimension (m-1) and canonical statistic $Z^*_{(1)} = B'_{(2)}X = (Z_{(1),2},\ldots,Z_{(1),m})'$.

Here is a famous log-linear model arising in genetics. Suppose a parent population contains alleles G,g at a certain locus, with frequency p,q = 1-p , respectively. Under the assumptions of random mating and no selection a generation of N offspring will have genotypes GG, Gg, gg according to a multinomial distribution with π given by

$$\text{(4)} \qquad \pi_1 = p^2 , \qquad \pi_2 = 2pq , \qquad \pi_3 = q^2 .$$

Such a multinomial distribution is called a *Hardy-Weinberg distribution*. This corresponds to a log-linear model with

$$\text{(5)} \qquad B = \begin{pmatrix} 1 & 2 \\ 1 & 1 \\ 1 & 0 \end{pmatrix} \qquad \theta_0 = \begin{pmatrix} 0 \\ \log 2 \\ 0 \end{pmatrix} .$$

Thus, $z_{(1)} = \binom{N}{2x_1 + x_2}$ is a sufficient statistic for the distributions of this log-linear family, and $z^*_{(1)} = 2x_1 + x_2$ is a minimal sufficient statistic.

(This log-linear family can be imbedded in a useful way in the original multinomial family as follows:

Let

$$\text{(6)} \qquad M = \begin{pmatrix} 2 & 1 & 0 \\ 0 & 1 & 2 \\ -1/3 & 2/3 & -1/3 \end{pmatrix} = \binom{M_1}{M_2} .$$

Then

$$M^{-1} = \begin{pmatrix} 5/12 & -1/12 & -1/2 \\ 1/6 & 1/6 & 1 \\ -1/12 & 5/12 & -1/2 \end{pmatrix} = (M_1^-, M_2^-) .$$

Let $\phi_0' = (0,0,-\ln 2)$ and $z_0' = (0,0,\frac{N}{3})$. According to Proposition 1.6 $Z = MX + z_0$ is the canonical statistic for an exponential family with corresponding canonical parameter $\phi = (M^{-1})'\theta + \phi_0$. In terms of the original variables $z_1 = 2x_1 + x_2$, $z_2 = 2x_3 + x_2$, $z_3 = x_2$, and $\phi_3 = (\frac{1}{2})\log(\pi_2^2/4\pi_1\pi_3)$, etc. The log-linear family described above is therefore the family of marginal distributions of (z_1, z_2) under the restriction $\phi_3 = 0$. The family of distributions corresponding to the restriction $\phi_3 = \phi_3^0 \neq 0$ also has a natural genetic interpretation as the distribution of a population after variable selection of genotypes. See Barndorff-Nielsen (1978, p.123); the generalization of this model to a multiallelic locus is also described there.)

REDUCTION TO A MINIMAL FAMILY

Any exponential family which is not minimal can be reduced to a minimal standard family through sufficiency, reparametrization, and proper choice of ν. This involves only a minor extension of the process used above in Proposition 1.5 and Theorem 1.7. This reduction is unique up to the appearance of linked affine transformations as in Proposition 1.6. Here are the details.

1.9 Theorem

Any k dimensional exponential family can be reduced by sufficiency, reparametrization, and proper choice of ν to an m dimensional minimal standard exponential family, for some $m \le k$. Let X,θ and Z,ϕ denote the canonical statistic and parameter for two such reductions to an m_1 and an m_2 dimensional minimal family, respectively. Then $m_1 = m_2$ and (X,θ), (Z,ϕ) are related as in 1.6(1).

Proof. The reduction to a minimal standard family will be performed in three steps. First, one may apply Proposition 1.5 to reduce to a standard k dimensional family.

Suppose for this family that $\dim \Theta = m' < k$. Thus $\Theta \subset \theta_0 + V$ where V is an m'-dimensional linear subspace. One may let P be the orthogonal projection on V and M_1, M_2 the corresponding orthonormal matrices described above in Theorem 1.7. Then $M_2\Theta = \phi_2^0$, a constant vector. By Theorem 1.7, $Z_1 = M_1X$ is sufficient, and its distributions form a standard exponential family, whose parameter space has dimension m'.

Thus it now suffices to consider the case of a standard m' dimensional exponential family whose parameter space also has dimension m' . Suppose for this family that dim $K = m < m'$. Then $K \subset x_0 + V$, similar to the previous situation. Let P be the orthogonal projection on V, and M_1, M_2 as above. Observe that

$$\text{(1)} \qquad \begin{aligned} \theta \cdot x &= \theta' M_1' M_1 x + \theta' M_2' M_2 x \\ &= \theta' M_1' M_1 x + \theta' M_2' M_2 x_0 \qquad \text{a.e.}\nu \quad . \end{aligned}$$

It follows that $Z_1 = M_1 X$ is a sufficient statistic whose distributions form a standard exponential family with natural parameter $M_1\theta$. (Actually Z is not merely sufficient, but is actually equivalent to X under ν.) Since $\dim (M_1 K) = \dim (M_1 \Theta) = m$ this family is the desired minimal family formed from the original family through reduction by sufficiency and reparametrization.

Suppose $\{p_\omega : \omega \in \Omega\}$ is a standard k dimensional exponential family relative to ν, and (X,θ), (Z,ϕ) denote the canonical statistics and parameters for two reductions of $\{p_\omega\}$ to a minimal standard exponential family. For the next step let $P^{(1)}_\theta$, $P^{(2)}_\phi$ denote their respective probability distributions with dimensions m_1 and m_2 respectively, etc.. Let $\omega_0 \in \Omega$. Since X and Z are each sufficient

$$\text{(4)} \qquad \frac{dP_\omega}{dP_{\omega_0}} = \frac{dP^{(1)}_{\theta(\omega)}}{dP^{(1)}_{\theta(\omega_0)}}(X(y)) = \frac{dP^{(2)}_{\phi(\omega)}}{dP^{(2)}_{\phi(\omega_0)}}(Z(y)) \qquad \text{a.e.}(\nu) \quad .$$

Now,

$$\begin{aligned} \frac{dP^{(1)}_{\theta(\omega)}}{dP^{(1)}_{\theta(\omega_0)}}(x) &= \frac{p^{(1)}_{\theta(\omega)}(x)}{p^{(1)}_{\theta(\omega_0)}(x)} \\ &= \exp(((\theta(\omega) - \theta(\omega_0)) \cdot x - (\psi^{(1)}(\theta(\omega)) - \psi^{(1)}(\theta(\omega_0))) \quad ; \end{aligned}$$

and similarly for $P^{(2)}$. Hence (4) yields

(5) $$(\theta(\omega) - \theta(\omega_0)) \cdot x(y) - U^{(1)}(\theta(\omega))$$

$$= (\phi(\omega) - \phi(\omega_0)) \cdot z(y) - U^{(2)}(\phi(\omega)) \qquad \text{a.e. } (\nu)$$

for all $\omega \in \Omega$.

Suppose $m = m_1 < m_2$. Since $\dim \{\phi(\omega) : \omega \in \Omega\} = m_2 > m$ there exist values $\alpha_i \in R$, $\omega_i \in \Omega$, $i=1,\ldots,m+1$, such that $0 = \sum_{i=1}^{m+1} \alpha_i(\theta(\omega_i) - \theta(\omega_0))$ and $\phi^* = \sum_{i=1}^{m+1} \alpha_i(\phi(\omega_i) - \phi(\omega_0)) \neq 0$. It follows from (5) that

(6) $$\phi^* \cdot z(y) = \text{const} \qquad \text{a.e. } (\nu) \quad .$$

But, (6) implies $K_2 \subset \{z : \phi^* \cdot z = \text{const}\}$ so that $\dim K_2 < m_2$. This contradicts the fact that the distributions of Z form a minimal standard family of dimension m_2. Hence $m_1 = m_2 = m$.

Now choose $\omega_1,\ldots,\omega_m$ so that $\{\theta(\omega_i) - \theta(\omega_0) : i=1,\ldots,m\}$ span R^m. The preceding argument shows that $\{\phi(\omega_i) - \phi(\omega_0) : i=1,\ldots,m\}$ must also span R^m. Let M, non-singular, be chosen so that

$$\phi(\omega_i) - \phi(\omega_0) = (M')^{-1}(\theta(\omega_i) - \theta(\omega_0)) \qquad i=1,\ldots,m \quad .$$

Then, as in 1.6(3),

(7) $$(\theta(\omega_i) - \theta(\omega_0)) \cdot x(y) - U(\theta(\omega_i))$$

$$= (\phi(\omega_i) - \phi(\omega_0)) \cdot Mx(y) - U(\phi(\omega_i))$$

$$= (\phi(\omega_i) - \phi(\omega_0)) \cdot z(y) - U(\phi(\omega_i)) \qquad \text{a.e. } (\nu) \quad .$$

Let $y_0 \in K$ be a value for which (7) is valid for $i=1,\ldots,m$. Then (7) yields

(8) $$(\phi(\omega_i) - \phi(\omega_0)) \cdot M(x(y) - x(y_0))$$

$$= (\phi(\omega_i) - \phi(\omega_0)) \cdot (z(y) - z(y_0)) \quad \text{a.e. } (\nu) \quad i=1,\ldots,m \quad .$$

This implies $M(x(y) - x(y_0)) = z(y) - z(y_0)$; which verifies 1.6(1) with $z_0 = z(y_0)$. ||

1.10 Definition

Let $\{p_\theta\}$ be a k-dimensional exponential family. Theorem 1.9 shows that there is a unique value, m, such that $\{p_\theta\}$ can be reduced to a minimal exponential family of dimension m. This value is called the *order* of the family p.

If $\{p_\theta\}$ is a standard family it is clear that its order m satisfies

(1) $$m \leq \min(\dim \Theta, \dim K) \quad .$$

In most cases equality holds in (1); however, it is possible to have inequality, even when $\{p_\theta\}$ is full.

In view of Theorem 1.9 there is no loss of generality in confining oneself to the study of minimal standard exponential families. A full minimal standard exponential family is also called a *canonical exponential family*.

RANDOM SAMPLES

A nearly trivial but very important application of the first part of Theorem 1.9 involves independent identically distributed (i.i.d.) observations from an exponential family.

1.11 Theorem

Let $X_1,\dots,X_n$ be i.i.d. observations from some k-dimensional standard exponential family with natural parameter space N and convex support K. Then $S = \sum_{i=1}^{n} X_i$ is a sufficient statistic. The distributions of S form a standard k-dimensional family with natural parameter space N and convex support $nK = \{s : \exists\, x \in K,\ s = nx\}$. The order of the families corresponding to S and to X_i are equal.

Proof: The joint density of $X_1,\dots,X_n$ with respect to $\nu\times \dots \times\nu$ is

$$p_\theta(x_1,\dots,x_n) = \exp\Big(\sum_{i=1}^{n} (\theta\cdot x_i - \psi(\theta))\Big)$$

$$= \exp\Big(\sum_{i=1}^{n} (\theta_i \cdot x_i - \psi(\theta_i))\Big) \quad \text{with} \quad \theta_i \equiv \theta \ .$$

Hence $X_1,\dots,X_n$ are canonical statistics from an nk-dimensional exponential family whose parameter space satisfies $\Theta = \{(\theta_1,\dots,\theta_n) \in R^{nk} : \theta_i \equiv \theta \in R^k\}$. Applying Theorem 1.7 yields that S is sufficient and comes from a standard k-dimensional family with natural parameter space N and convex support nK. (All this is also obvious from the fact that

$$p_\theta(x_1,\dots,x_n) = \exp\Big(\theta \sum_{i=1}^{n} x_i - n\psi(\theta)\Big) \quad .)$$

It is easily checked that any linear map which transforms the distributions of X_i to a minimal family also transforms those of S to one, and conversely. This yields the assertion concerning the order of the families corresponding to S and X_i . ||

Note that the cumulant generating function for the exponential family generated by S is

(1) $$n\psi(\theta) \ .$$

The sufficient statistic $\bar{X} = n^{-1}S$ also has distributions from an exponential family. (Apply Theorem 1.6.) Here, the natural parameter space is nN and the convex support is K. The cumulant generating function for $\bar{X}$ corresponding to the point $\phi = n\theta$ in its natural parameter space is

(2) $$n\psi(\phi/n) \quad .$$

(Under appropriate additional conditions a family of distributions for which there is a nontrivial sufficient statistic based on a sample of size n must be an exponential family. See Dynkin (1951) and Hipp (1974).)

1.12 Examples

Example 1.2 displays an instance of this theorem. If Y is normal with mean μ and variance σ^2 then $X = (Y, Y^2)$ is the canonical statistic of a minimal standard exponential family having canonical parameter $\theta = (\mu/\sigma^2, -1/2\sigma^2)$. Thus if one has i.i.d. observations $Y_1,\ldots,Y_n$ then $S = \sum_{i=1}^{n} X_i = (\sum_{i=1}^{n} Y_i, \sum_{i=1}^{n} Y_i^2)$ is a sufficient statistic; and its distributions form a minimal standard exponential family.

As another example, suppose Y is a member of the *gamma family* with unknown index, α, and scale, σ. The density of Y relative to Lebesque measure on $(0, \infty)$ is

(1) $$f(y) = (\sigma^{\alpha}\,\Gamma(\alpha))^{-1} y^{(\alpha-1)}\, e^{-y/\sigma} \,, \qquad y > 0 \quad .$$

We will use the notation $Y \sim \Gamma(\alpha, \sigma)$. Note that $\Gamma(m/2, 2) = \chi^2_m$. These distributions form a two-dimensional exponential family with canonical statistic $(Y, \ln Y)$ and canonical parameters $(-1/\sigma, \alpha)$. If $Y_1,\ldots,Y_n$ are i.i.d. with density (1) then $S_1 = \sum_{i=1}^{n} Y_i$ and $S_2 = \sum_{i=1}^{n} \ln Y_i$ form a two-dimensional exponential family. It is interesting to note that the marginal distribution of S_1/n also has a density of the form (1) with index $n\alpha$ and scale $n\sigma$. (Here, as well as in the preceding normal example, S_1 is strongly reproductive in the terminology of Barndorff-Nielsen and Blaesild (1983b). For more details see Theorem 2.14 and Example 2.15.)

Another example of interest is provided by the Poisson distribution; where Y has probability function

(2) $$\Pr\{Y = y\} = \lambda^y e^{-\lambda}/y! \qquad y=0,1,\ldots \quad .$$

We will use the notation $Y \sim P(\lambda)$. Then $X = Y$ comes from a one-dimensional exponential family with canonical parameter $\theta = \ln \lambda$. The distribution of $S = \sum_{i=1}^{n} Y_i$ is itself Poisson with natural parameter $\theta + \ln n = \ln n\lambda$.

CONVEXITY PROPERTY

Here is an important fundamental fact about exponential families.

1.13 Theorem

(i) N is a convex set and ψ is convex on N.

(ii) ψ is lower semi-continuous on R^k and is continuous on $N°$.

(iii) $P_{\theta_1} = P_{\theta_2}$ if and only if

(1) $$\psi(\alpha\theta_1 + (1 - \alpha)\theta_2) = \alpha\psi(\theta_1) + (1 - \alpha)\psi(\theta_2)$$

for some $0 < \alpha < 1$. In this case (1) is then valid for all $0 \leq \alpha \leq 1$.

(iv) If $\dim K = k$ (in particular, if $\{p_\theta\}$ is minimal) then ψ is strictly convex on N, and $P_{\theta_1} \neq P_{\theta_2}$ for any $\theta_1 \neq \theta_2 \in N$.

Proof: Let $\theta_1, \theta_2 \in N$, $0 < \alpha < 1$. Then by Hölder's inequality

(2) $$\begin{aligned} \exp(\psi(\alpha\theta_1 + (1 - \alpha)\theta_2)) &= \int \exp((\alpha\theta_1 + (1 - \alpha)\theta_2) \cdot x)\nu(dx) \\ &= \int (\exp \theta_i \cdot x)^\alpha \cdot (\exp \theta_2 \cdot x)^{(1-\alpha)} \nu(dx) \\ &\leq \left(\int \exp(\theta_1 \cdot x)\nu(dx)\right)^\alpha \left(\int \exp(\theta_2 \cdot x)\nu(dx)\right)^{(1-\alpha)} \\ &= \exp(\alpha\psi(\theta_1) + (1 - \alpha)\psi(\theta_2)) \quad . \end{aligned}$$

This proves the convexity of ψ, and the convexity of N follows easily.

There is strict inequality in (1) unless

(3) $$\theta_1 \cdot x = \theta_2 \cdot x + K \qquad \text{(a.e. } (\nu))$$

for some constant K; in which case there is equality. (3) is equivalent to

$e^{\theta_1 \cdot x} = e^K e^{\theta_2 \cdot x}$ a.e.(ν) which is equivalent to the assertion $P_{\theta_1} = P_{\theta_2}$.

If (3) holds for some $\theta_1 \neq \theta_2$ then dim $K \leq k - 1$. Hence dim $K = k$ implies $P_{\theta_1} \neq P_{\theta_2}$ for any $\theta_1 \neq \theta_2 \in N$.

Finally, for the continuity assertions, note first that $\lambda(\theta) = \int \exp(\theta \cdot x)\nu(dx)$ is lower semi-continuous by Fatou's lemma. Hence ψ is lower semi-continuous. Any convex function defined and finite on a convex set N of R^k must be continuous on $N°$. (We leave this as an exercise on convex sets.) ||

Be careful about the above result -- the fact that ψ is strictly convex on N does not imply that N is strictly convex; for a simple example, see Example 1.2 which involves a minimal family for which

$$N = \{(\theta_1, \theta_2) : \theta_1 \in R, \ \theta_2 < 0\} \quad .$$

Usually ψ is continuous on all of N. However examples can be constructed when $k \geq 2$ where this is not the case.

This simple theorem has an interesting direct application.

1.14 Example

Let Y be m-variate normal with mean μ and covariance matrix Σ. We will use the notation $Y \sim N(\mu, \Sigma)$. Also, $\delta_{ij} = 1$ if i=j and = 0 if i≠j . The density of Y with respect to Lebesgue measure is

$$(1) \qquad \phi_{\mu,\Sigma}(y) = (2\pi)^{-m/2}|\Sigma|^{-\frac{1}{2}}\exp(\mathrm{tr}(-\Sigma^{-1}(y - \mu)(y - \mu)'/2))$$

$$= (2\pi)^{-m/2}|\Sigma|^{-\frac{1}{2}}\exp((\Sigma^{-1}\mu) \cdot y + \mathrm{tr}((-\Sigma^{-1}/2)(yy')) - \mu'\Sigma^{-1}\mu/2) \ .$$

It follows that the distributions of Y form an $(m + m(m+1)/2)$ dimensional exponential family with canonical statistics $Y_1,\ldots,Y_m$, $\{Y_iY_j/(1 + \delta_{ij}) : i \leq j\}$ and corresponding canonical parameters $(\Sigma^{-1}\mu)_1,\ldots,(\Sigma^{-1}\mu)_m$, $\{(-\Sigma^{-1})_{ij} : i \leq j\}$. For the following it is convenient to label these statistics $X_1,\ldots,X_m$, $\{X_{ij} : i \leq j\}$ and the corresponding parameters as $(\theta_1,\ldots,\theta_m, \{\theta_{ij} : i \leq j\})$. Write $\theta = (\theta_1,\ldots,\theta_m)$, $\mathcal{Q} = (\theta_{ij})$. Ignoring the factor $(2\pi)^{-m/2}$, which can be absorbed into the measure ν, the cumulant generating function is

(2) $\psi(\cdot) = (-\tfrac{1}{2})\log|\Sigma^{-1}| + (\mu'\Sigma^{-1}\mu)/2 = (-\tfrac{1}{2})\log(|-\mathcal{Q}|) - \theta'\mathcal{Q}\theta/2$.

Note that $N = \{(\theta, \{\theta_{ij} : i \le j\}) : -\mathcal{Q}$ is positive definite$\}$. It is easy to check that N is open, so that this family is regular. By Theorem 1.12

(3) $$\psi(0, \{\theta_{ij} : i \le j\}) = (-\tfrac{1}{2})\log(|-\mathcal{Q}|)$$

is strictly convex in the variables $\{\theta_{ij} : i \le j\}$ over the set where $\mathcal{Q}$ is positive definite. To reinterpret this result slightly, let $B = -\mathcal{Q}$; then (3) yields that

(4) $$\log |B| \quad \text{is strictly concave}$$

as a function of the variables $\{b_{ij} : i \le j\}$ over the region where B is positive definite. (4) yields

(5) $$|B^{-1}| = |B|^{-1} \quad \text{is strictly convex} \quad .$$

((4) can also be proven by directly calculating $\dfrac{\partial^2}{\partial b_{ij}\,\partial b_{k\ell}} \log|B|$, and showing the resulting $\binom{k+1}{2} \times \binom{k+1}{2}$ matrix is positive definite. The above proof is much simpler !)

CONDITIONAL DISTRIBUTIONS

Let ν be a given σ-finite measure on the Borel subsets of R^k, and $P \ll \nu$ a probability measure with density p. Assume (without loss of generality) that $0 \in N$ so that ν is finite. Let $M_1 : R^k \to R^m$ be linear, $M_1(x) = M_1 x$. Then the conditional measure of ν given $z_1 = M_1(X)$ exists. It will be denoted by $\nu(\cdot|M_1 X = z_1)$ or $\nu(\cdot|z_1)$. The conditional distribution of P given $M_1(X)$ exists and has density proportional to $p(\cdot)$ relative to $\nu(\cdot|z_1)$ over the set $\{x : M_1(X) = z_1\}$. (More generally these facts are true if M_1 is any Borel measurable function. See, for example, Neveu (1965).)

The above situation resembles that described in 1.7. Let $M_2 : R^k \to R^{k-m}$ be an orthogonal complement of M_1. Then

$$M_2 : \{x : M_1(x) = z_1\} \to R^{k-m}$$

is 1 - 1.

We will also use the symbol $\nu(\cdot|z_1)$ for the equivalent conditional distribution of $M_2(X)$ given $M_1(X) = z_1$. As before,

$$\phi = M'^{-1}\theta = \binom{M_1'^{-}}{M_2'^{-}}\theta = \binom{\phi_1}{\phi_2} .$$

It is always possible to choose M_2 to be "orthonormal" so that

$$M_2^{-} = M_2' , \quad \text{and so} \quad M_2'^{-} = M_2 .$$

To do so simplifies somewhat the resulting formulae.

<u>1.15 Theorem</u>

The distribution of $Z_2 = M_2X$ given $Z_1 = M_1X$ depends only on $\phi_{(2)} = M_2'^{-}\theta$. For fixed $Z_1 = z_1$ these distributions form the (k-m) dimensional exponential family generated by the measure defined by $\nu(\cdot|z_1)$.

Let N_{z_1} denote the natural parameter space of this conditional family. Then $\phi_2 \in M_2'^{-}N$ implies

(1) $$\phi_2 \in N_{M_1X} \quad \text{a.e.}(\nu) .$$

Furthermore, if $\{p_\theta\}$ is regular then

(2) $$M_2'^{-}N \subset N_{M_1X} \quad \text{a.e.}(\nu) .$$

Proof: The conditional density of Z_2 given $Z_1 = z_1$ is proportional to

$$p_\theta((z_1, z_2)) = \exp(\phi_1 \cdot z_1 + \phi_2 \cdot z_2 - \psi(\theta)) .$$

Hence the density of Z_2 given $Z_1 = z_1$ relative to $\nu(\cdot|z_1)$ can be written as

(3) $$p_\phi(z_2) = \exp(\phi_2 \cdot z_2 - \psi_z(\phi_2))$$

where

(4) $$\psi_{z_1}(\phi_2) = \ln(\int \exp(\phi_2 \cdot z_2)\nu(dz_2|z_1)) .$$

The natural parameter space N_{z_1} is the set$\{\phi_2\}$, for which the integral on the right of (4) is finite. Let $\phi_2 \in M_2'^{-}N$. There is thus a $\theta \in N$

for which $\phi_2 = M_2'^{-}\theta$. Let ν^* denote the marginal measure on R^m defined by $\nu^*(A) = \nu(M_1^{-1}(A))$. Then

$$\infty > \int \exp(\theta \cdot x)\nu(dx) = \int\{\int \exp(\phi_1 \cdot z_1 + \phi_2 \cdot z_2)\nu(dz_2|z_1)\}\nu^*(dz_1) \quad .$$

Hence

$$\infty > \int \exp(\phi_2 \cdot z_2)\nu(dz_2|z_1)$$

for almost every $z_1(\nu^*)$. This verifies (1).

Suppose $\{p_\theta\}$ is regular. Let $\{\theta_i: i=1,\ldots,\} \subset N$ be a countable dense subset of N. $\{M_2'^{-}\theta_i : i=1,\ldots\}$ is dense in $M_2'^{-}N$. $M_2'^{-}$ is a linear map. Hence $M_2'^{-}N$ is convex and open since N is convex (by Theorem 1.13) and open (by assumption). It follows that

$$\text{(5)} \qquad \text{conhull } \{M_2'^{-}\theta_i : i=1,\ldots\} = M_2'^{-}N \quad .$$

(We leave (5) as an exercise on convex sets.)

Since $\{\theta_i\}$ is countable it follows from (1) that

$$M_2'^{-}\theta_i \subset N_{M_1X} \quad \text{for all} \quad i=1,\ldots, \quad \text{a.e.}(\nu) \quad .$$

Thus

$$M_2'^{-}N = \text{conhull } \{M_2'^{-}\theta_i : i=1,\ldots\} \subset N_{M_1X} \quad \text{a.e.}(\nu) ,$$

since N_{M_1X} is convex; which proves (2). ||

The above result can be given an alternate interpretation under which the conditional distributions of X given $X \in L$ form an exponential family, for L a given linear variety in R^k. See 1.7(8). We omit the details.

Here are two important simple applications of the above ideas.

<u>1.16 Example</u>

Let $X_1,\ldots,X_k$ be independent Poisson variables with expectations λ_i . See 1.12(2). Then $X = (X_1,\ldots,X_k)$ is the canonical statistic of a standard exponential family with natural parameter θ: $\theta_i = \ln \lambda_i \quad i=1,\ldots,k$. The dominating measure has $\nu(\{x\}) = 1/\prod_{i=1}^{k} x_i!$. Let $N > 0$ be an integer.

Then the distributions of X given $\sum_{i=1}^{k} X_i = N$ form a standard exponential family with dominating measure

(1) $$\nu(\{x\}|\sum_{i=1}^{k} x_i = N) = 1/\prod_{i=1}^{n} x_i!, \quad \text{for} \quad \sum_{i=1}^{k} x_i = N .$$

This measure is proportional to the measure 1.3(1) which generates the multinomial distribution. Hence the conditional distribution is multinomial (N,π).

The value of π can be easily computed as follows: orthogonally project onto $\{\theta: \Sigma\theta_i = 0\}$ which is the linear subspace parallel to $\{x : \Sigma x_i = N\}$. This yields $(\theta - \bar{\theta}1)$ (where $\bar{\theta} = k^{-1}\sum_{i=1}^{k}\theta_i$) as the natural parameter of the conditional multinomial distribution. Thus

$$\pi_i = ce^{\theta_i - \bar{\theta}}$$

with $c = (\sum_{i=1}^{k} e^{\theta_i - \bar{\theta}})^{-1}$. Substituting $\theta_i = \ln \lambda_i$ yields

(2) $$\pi_i = \lambda_i / \sum_{i=1}^{k} \lambda_i .$$

1.17 Example

Let X be k-variate normal with mean μ and covariance Σ. For Σ given the distributions of X form a standard exponential family with natural parameter $\theta = \Sigma^{-1}\mu$. (This can easily be checked directly or derived from Example 1.14 by using Theorem 1.7.) The dominating measure for this family is proportional to $\nu(dx) = \exp(-x'\Sigma^{-1}x/2)dx$.

Let $z_1 = (x_1,\ldots,x_m)$, $z_2 = (x_{m+1},\ldots,x_k)$. The conditional distributions of Z_2 given $Z_1 = z_1$ form an exponential family. The natural parameter for this family is just $\phi_2 = (\theta_{m+1},\ldots,\theta_k)'$.

Partition Σ as

(1) $$\Sigma = \begin{pmatrix} \Sigma_{11} & \Sigma_{12} \\ \Sigma_{21} & \Sigma_{22} \end{pmatrix} \quad \text{with} \quad \Sigma_{11}(m \times m) , \text{ etc.}$$

Then

(2) $$\Sigma^{-1} = \begin{pmatrix} (\Sigma_{11} - \Sigma_{12}\Sigma_{22}^{-1}\Sigma_{21})^{-1} & -\Sigma_{11}^{-1}\Sigma_{12}(\Sigma_{22}-\Sigma_{21}\Sigma_{11}^{-1}\Sigma_{12})^{-1} \\ -(\Sigma_{22}-\Sigma_{21}\Sigma_{11}^{-1}\Sigma_{12})^{-1}\Sigma_{21}\Sigma_{11}^{-1} & (\Sigma_{22}-\Sigma_{21}\Sigma_{11}^{-1}\Sigma_{12})^{-1} \end{pmatrix}$$

$$= \begin{pmatrix} \Sigma^{11} & \Sigma^{12} \\ \Sigma^{21} & \Sigma^{22} \end{pmatrix}, \text{ say.}$$

((2) is a general formula for block symmetric positive definite matrices. Note that $\Sigma^{12} = -\Sigma_{11}^{-1}\Sigma_{12}(\Sigma_{22} - \Sigma_{21}\Sigma_{11}^{-1}\Sigma_{12})^{-1} = -\Sigma_{22}^{-1}\Sigma_{21}(\Sigma_{11} - \Sigma_{12}\Sigma_{22}^{-1}\Sigma_{21})^{-1}$.) Note that the natural parameter can be written as

$$\phi_2 = \begin{pmatrix} \theta_{m+1} \\ \vdots \\ \theta_k \end{pmatrix} = (\Sigma^{-1}\mu)_2 = \Sigma^{21}\mu_{(1)} + \Sigma^{22}\mu_{(2)}$$

where

$$\mu = \begin{pmatrix} \mu_{(1)} \\ \mu_{(2)} \end{pmatrix} .$$

Consider the case where $z_1 = 0$. The conditional dominating measure is

$$\nu(dz_2|0) = c \exp(-z_2'\Sigma^{22} z_2/2)$$

and is thus a normal density with mean 0, variance-covariance $(\Sigma^{22})^{-1} = \Sigma_{22} - \Sigma_{21}\Sigma_{11}^{-1}\Sigma_{12} = \Sigma^*$, say. It follows that the conditional density of Z_2 given $Z_1 = 0$ is normal with this covariance matrix and with mean μ^* given by

$$\Sigma^{*-1}\mu^* = \phi_2 ,$$

since ϕ_2 must be the value of the natural parameter for both the unconditional and conditional family. Hence

(3) $$\mu^* = \Sigma^*\phi_2 = \Sigma^*(\Sigma^{21}\mu_{(1)} + \Sigma^{22}\mu_{(2)}) = -\Sigma_{21}\Sigma_{11}^{-1}\mu_{(1)} + \mu_{(2)} .$$

For $z_1 \neq 0$ it is convenient to use the location invariance of the normal family. The conditional distribution under (μ,Σ) of $Z_{(2)}$ given $Z_{(1)} = z_{(1)}$ is the same as the conditional distribution under $(\begin{pmatrix}\mu_{(1)} - z_{(1)} \\ \mu_{(2)}\end{pmatrix}, \Sigma)$ of $Z_{(2)}$ given $Z_{(1)} = 0$. By the preceding this is normal with covariance matrix $\Sigma^* = (\Sigma^{22})^{-1}$ and mean $\mu_{(2)} - \Sigma_{21}\Sigma_{11}^{-1}(\mu_{(1)} - z_{(1)})$.

EXERCISES

1.1.1 (a) Let C be any closed convex set in R^k. Show that there exists a standard exponential family with $N = C$. [$C = \bigcap_{i=1}^{\infty} \{\theta: v_i \cdot \theta \le c_i\}$ with $||v_i|| = 1$. Let ν_i denote Lebesgue measure on the ray $\{x: x = \alpha v_i, \alpha > 0\}$ and let $\nu = \sum_{i=1}^{\infty} 2^{-i} \exp(c_i v_i \cdot x)\nu_i/(1+||x||^2)$. The result is also true, but harder to prove, if C is an open convex set.]

(b) Let $C = \{(\theta_1, \theta_2): ||\theta||^2 < 1\} \cup \{(0, 1)\}$ and show there exists an exponential family with $N = C$.

1.2.1 Verify 1.2(5) (including the formula for ν which precedes it). Note that when $n = 1$ the measure ν can be described by the relations $x_2 = x_1^2$ and $\nu(dx_1) = dx_1/\sqrt{2\pi}$.

1.7.1 (i) Let $Z = MX$ as in Theorem 1.7. Show that Z_1 is independent of Z_2 for some $\theta \in \Theta$ if and only if Z_1 is independent of Z_2 for all $\theta \in \Theta$.

(ii) Give an example to show that the assertion is false if Z_1, Z_2 are non-linear transformations of X. [(i) Assume independence at $\theta = 0$. (ii) Let X be bivariate normal with mean μ and covariance I, and $Z_1 = ||x||$, $Z_2 = \tan^{-1}(x_2/x_1)$.]

1.7.2 Consider the situation of Theorem 1.7. Suppose the original family $\{p_\theta: \theta \in N\}$ is full and minimal. Then the family of distributions of Z_1 for $\phi_1 \in \Phi_{\phi_2^0}$ is full. It is minimal if and only if there is a $\theta \in \operatorname{int} N$ with $M_2'\phi_2^0 = \theta$. [For a situation where the family of distributions of Z_1 is not minimal use Exercise 1.1.1(b), let M be as in (4), and let $\phi_2^0 = 1$.]

1.7.3 (a) Show that if 1.7(7) or (8) are satisfied then the distributions of $Z_1 = M_1X$ form a standard exponential family with natural parameter ϕ_1.

(b) Give an example to show that the distributions of $Z_1 = M_1X$ may form a standard exponential family with natural parameter different from ϕ_1 even when 1.7(7) and (8) fail. [Consider the distribution of X_1 when X is multinomial of dimension $k \geq 3$, or equivalently, of X_1^* with X^* as in Example 1.3. There are also some other interesting instances of this phenomenon.]

<u>1.8.1</u> (Contingency table under independence). Consider a 2×2 contingency table in which the observations are Y_{ij}, $1 \leq i, j \leq 2$, and have a multinomial (N, p) distribution with $p = \{p_{ij}, 1 \leq i, j \leq 2\}$. Under the model of independence $p_{ij} = p_{i+} \cdot p_{+j}$ where $p_{i+} = \sum_j p_{ij}$, etc.. Write this independence model as a log-linear model in a fashion so that the coordinates of the natural (minimal) sufficient statistic are independent binomial variables. Generalize to the model of independence in an r×c contingency table. (For further log-linear models in contingency tables, see Haberman (1974), (1979).)

<u>1.10.1</u> Show that in any standard exponential family of dimension k and order m, $m + k \geq \dim K + \dim \Theta$. Give an example in which $m < \min(\dim \Theta, \dim K)$. [The simplest example has $m = 0$, $\dim \Theta = \dim K = 1$, $k = 2$.]

<u>1.12.1</u> From many points of view the negative binomial distributions are the discrete analog to the gamma distributions. The *negative binomial*, $N\mathcal{B}(\alpha, p)$, distribution has probability function

$$P(x) = \frac{\Gamma(x + \alpha)}{\Gamma(x+1)\Gamma(\alpha)} (1 - p)^{\alpha} p^{x}, \quad x=0,1,\ldots \quad .$$

Show that for fixed α the family $N\mathcal{B}(\alpha, \cdot)$ is a one parameter exponential family, but that -- unlike the $\Gamma(\alpha, \sigma)$ situation -- the family $N\mathcal{B}(\alpha, p)$ $\alpha > 0$, $0 < p < 1$ is not an exponential family.

<u>1.12.2</u> Let ν denote counting measure on $\{(0,0), (1,1), (2,0), (3,1), (4,0), \ldots\} \subset R^k$. Show that the exponential family generated by ν has the following properties: X_1 has a *geometric distribution*, $Ge(p_1) = N\mathcal{B}(1,p)$; X_2 has a binomial distribution, $\mathcal{B}(p_2)$; $(X_1 - X_2)/2$ has a geometric distribution $Ge(p_3)$ and $(X_1 - X_2)/2$ is independent of X_2. Write p_1, p_2, p_3 in terms of the natural parameters θ_1, θ_2.

1.12.3 Let $Z_1,\ldots,Z_m$ be i.i.d. $N(\mu,\sigma^2)$. Let $X = \sum_{i=1}^{m} Z_i^2$. Then X has a scaled non-central χ^2 distribution with m degrees of freedom, non-centrality parameter $\delta = m\mu^2/\sigma^2$, and scale parameter σ^2. Denote this distribution by $\chi_m^2(\delta, \sigma^2)$. (i) This distribution has density

$$(1) \qquad g(x) = \sum_{k=0}^{\infty} \frac{\lambda^k e^{-\lambda}}{k!} \frac{(\frac{x}{\sigma^2})^{(m/2)+k-1} e^{-x/(2\sigma^2)}}{\sigma^2 \Gamma(k+\frac{m}{2})\, 2^{k+m/2}}, \qquad x > 0,$$

where $\lambda = \delta/2$. (From the form of (1) it is evident that $K = k \sim P(\lambda)$ and $X|K \sim \Gamma(k + m/2, \sigma^2)$; thus (X/σ^2) K is central χ^2 with $k + \frac{m}{2}$ degrees of freedom.) (ii) The distributions of X can also be represented as the marginal distribution of X_1 from a canonical two parameter exponential family generated by a measure ν supported on $\{(x_1, x_2)\colon x_1 > 0,\ x_2 = 0,1,\ldots\}$. [(i) By change of variables and series expansion prove (1) for the case m=1. (1) for general m then follows from facts about sums of Poisson and gamma variables. (ii) Let ν be the measure generated by (1) with $\sigma^2 = 1$, $\lambda = 1$.]

1.13.1 (i) Show that when k = 1 then ψ must be continuous on N. [Use 1.13 and convexity of N.]

(ii) More generally, let $\theta_0,\theta_1 \in N$ and $\theta_\rho = (1-\rho)\theta_0 + \rho\theta_1$ and show $\psi(\theta_\rho)$ is continuous in ρ for $0 \le \rho \le 1$. [Reason as in (i), or use Theorem 1.7 and (i).]

(iii) Give an example of an exponential family in which ψ is not continuous on N. [Exercise 1.1.1(b) provides an example.]

1.13.2 Generalize 1.13.1(ii) as follows: let $\theta \in N$ and $\theta_j \in N$, j=1,...,J. Let $\tau_i \in \text{conhull}\{\theta, \theta_1,\ldots,\theta_J\}$, i=1,... and $\tau_i \to \theta$. Then $\psi(\tau_i) \to \psi(\theta)$. [Write $\tau_i = \sum_j \alpha_{ij}[(1-\rho_i)\theta_j + \rho_i\theta]$ with $\alpha_{ij} \ge 0$, $\sum_j \alpha_{ij} = 1$, and $\rho_i \uparrow 1$. Use 1.13.1(ii) and the fact that $\psi(\tau_i) \le \sum_j \alpha_{ij}\, \psi((1-\rho_i)\theta_j + \rho_i\theta)$.]

1.13.3 Let $Y = (Y_0 = 1, Y_1,\ldots Y_n)'$ be the initial state and n further observations from an S-state Markov chain with transition matrix P (i.e., $P(Y_\ell = j|Y_{\ell-1} = i) = p_{ij}$, $1 \le i,j \le S$, $\ell=1,\ldots,n$). Let N denote the sample transition matrix, $N = \{n_{ij}\}$,

(1) $$n_{ij} = \sum_{\ell=1}^{n} \chi_{\{i,j\}}(Y_{\ell-1}, Y_\ell) .$$

Suppose $p_{ij} > 0$, $1 \leq i,j \leq S$. Show that the distributions of Y form an S^2 dimensional exponential family with canonical statistic $N = \{n_{ij}\}$ and canonical parameters $\{\log p_{ij}\}$. Show that if $n \geq 3$ the family has order $S^2 - 1$. [Let E_{ij} denote the matrix with i,j-th entry 1 and all other entries 0. Show that for given $1 \leq i,j \leq K$ there exist sample points N_1, N_2 having positive probability and that $N_1 + (E_{ii} - E_{ji}) = N_2$ and (other) points N_1, N_2 such that $N_1 + (E_{ii} - E_{11}) = N_2$.]

1.14.1 Univariate General Linear Model (G.L.M.). Let Y be m-variate normal, $Y \sim N(\mu, \sigma^2 I)$, $\mu \in R^m$, $\sigma^2 > 0$. (a) Show that this is an m+1 dimensional exponential family. (b) In the G.L.M. μ is restricted by

$$\mu = B\beta , \qquad \beta \in R^r$$

with B a known m×r matrix. Assume (for convenience) B has rank r. Show that this is a full (r+1) dimensional exponential family. [Use Example 1.14 and Theorem 1.7.]

1.14.2 Matrix normal distribution. Let $\mu = \{\mu_{ij}\}$ be an m×q matrix and let $\Gamma = \{\gamma_{ij}\}$ and $\Sigma = \{\sigma_{ij}\}$ be m×m and q×q positive definite matrices, respectively. Let $Y = \{Y_{ij}\}$ be an m×q random matrix whose entries have a multivariate normal distribution with

$$E\, Y_{ij} = \mu_{ij} \qquad \operatorname{Cov} Y_{ij} Y_{i'j'} = \gamma_{ii'}\sigma_{jj'} .$$

This is the matrix normal distribution, denoted by $Y \sim N(\mu, \Gamma, \Sigma)$.

(a) Show that Y has density (relative to Lebesgue measure on R^{mq})

$$f(y) = (2\pi)^{-mq/2} |\Gamma|^{-m/2} |\Sigma|^{-q/2} \exp \operatorname{tr} (-\Gamma^{-1}(Y - \mu)\Sigma^{-1}(Y - \mu)'/2) .$$

[See Arnold (1981, Theorem 17.4).]

(b) Reduce this to an $mq + \frac{m(m+1)q(q+1)}{4}$ dimensional minimal exponential family with canonical parameters $\theta_{ij} = \Gamma^{-1}\mu\Sigma^{-1}$, $1 \leq i \leq m$, $1 \leq j \leq q$, and $\theta_{ii'jj'} = \gamma^{ii'}\sigma^{jj'}$, $1 \leq i \leq i' \leq m$, $1 \leq j \leq j' \leq q$, where $\Gamma^{-1} = \{\gamma^{ij}\}$,

$\Sigma^{-1} = \{\sigma^{ij}\}$.

(c) Show that if $m \geq 2$ and $q \geq 2$ this is not a full exponential family. Rather, Θ is an $mq + m(m+1)/2 + q(q+1)/2 - 1$ dimensional differentiable manifold inside of N.

(An alternate notation involves writing $Y = (Y_{(1)}, \ldots, Y_{(q)})$ and defining $(\text{vec } Y)' = (Y'_{(1)}, \ldots, Y'_{(q)})$. Then $Y \sim N(\mu, \Gamma, \Sigma)$ is the same as $\text{vec } Y \sim N(\text{vec } \mu, \Sigma \otimes \Gamma)$ where $\otimes$ denotes the Kronecker product.

1.14.3 Multivariate Linear Model (M.L.M.). Here $Y \sim N(\mu, I, \Sigma)$ with Σ positive definite and

$$\mu = B\beta$$

with B a known $m \times r$ matrix and β an $(r \times q)$ matrix of parameters. Assume (for convenience) B has rank r. Show that this can be reduced to a full minimal regular exponential family of dimension $rq + q(q+1)/2$.

1.14.4 Wishart distribution. Let $X = (x_{ij})$ and $\Sigma = (\sigma_{ij})$ be symmetric $m \times m$ positive definite matrices. The matrix $\Gamma(\alpha, \Sigma)$ distribution has density

$$(1) \qquad p_{\alpha,\Sigma}(X) = \frac{|X|^{\alpha-(m+1)/2} \exp\text{tr}(-\Sigma^{-1}X)}{\Gamma_m(\alpha)|\Sigma|^{\alpha}}$$

where

$$\Gamma_m(\alpha) = \pi^{m(m-1)/4} \prod_{i=1}^{m} \Gamma(\alpha - (i-1)/2) , \quad \alpha > (m-1)/2 .$$

Show this is an exponential family, and describe the natural observations, natural parameters, and cumulant generating function.

(If Y_i, $i=1,\ldots,n$, are independent $N(0, \Sigma)$ vectors then $\sum_{i=1}^{n} Y_i Y_i' = X$ has the $\Gamma(\frac{n}{2}, 2\Sigma)$ distribution. This is also called the *Wishart* (n, Σ) distribution and denoted by $W(n, \Sigma)$. See e.g. Arnold (1981). Also $\sum_{i=1}^{n} (Y_i - \bar{Y})(Y_i - \bar{Y})' \sim W(n-1, \Sigma)$.)

1.15.1 Consider a 2×2 contingency table (see Exercise 1.8.1). Find the conditional distribution of Y_{ij} given $Y_{i+} = \sum_{j=1}^{2} Y_{ij}$ and $Y_{+j} = \sum_{i=1}^{2} Y_{ij}$. Show that these conditional distributions depend only on the given values Y_{i+}, Y_{+j} and on the *odds ratio* $p_{11}p_{22}/p_{12}p_{21}$ and form a one-parameter exponential family. [Under the independence model the distribution is hypergeometric and independent of p.]

Chapter 2. Analytic Properties

DIFFERENTIABILITY AND MOMENTS

The cumulant generating function has several nice properties. Among these are the fact that its defining expression may be differentiated under the integral sign. In this manner one obtains the moments of X from the derivatives of ψ.

One needs first to establish a simple bound.

2.1 Lemma

Let $B = \text{conhull}\,\{b_i : i=1,\dots,I\} \subset R^k$. Let $C \subset B°$ be compact and let $b_0 \in C$. Then there are constants K_ℓ (depending on C,B) $\ell=0,1,\dots$ such that

$$(1) \qquad ||x||^\ell e^{b\cdot x} \le K_\ell \sum_{i=1}^{I} e^{b_i\cdot x} \qquad \forall\; b \in C, \quad x \in R^k .$$

Also,

$$(2) \qquad \left| \frac{e^{b\cdot x} - e^{b_0\cdot x}}{||b - b_0||} \right| \le K_1 \sum_{i=1}^{I} e^{b_i\cdot x}, \quad b \in C, \quad x \in R^k .$$

Proof. Let $\varepsilon > 0$. Note that there exists a $K_{\ell,\varepsilon} < \infty$ such that

$$|r|^\ell \le K_{\ell,\varepsilon}\, e^{\varepsilon|r|} \qquad \forall \quad r \in R$$

since

$$\lim_{|r|\to\infty} |r|^\ell/e^{\varepsilon|r|} = 0 .$$

Let $\{e_i : i=1,\dots,k\}$ denote the elementary (orthogonal) unit vectors in R^k. Then

$$||x||^\ell \le k^{(\ell-2)/2} \sum_{i=1}^{k} |x_i|^\ell \le K'_{\ell,\varepsilon} \sum_{i=1}^{k} e^{\varepsilon|x_i|} < K'_{\ell,\varepsilon} \sum_{i=1}^{k} (e^{\varepsilon e_i\cdot x} + e^{-\varepsilon e_i\cdot x}),$$

where $K'_{\ell,\varepsilon} = k^{(\ell-2)/2} K_{\ell,\varepsilon}$. Choose $\varepsilon > 0$ such that $(b \pm \varepsilon e_i) \in B$, $i=1,\ldots,k$, for all $b \in C$. See Figure 2.1(1). By convexity

$$e^{(b\pm\varepsilon e_i)\cdot x} \leq \max(e^{b_i\cdot x}) \quad ,$$

since $e^{a\cdot x}$ is convex in $a \in R^k$ and $(b\pm\varepsilon e_i) \in B = \text{conhull}\ \{b_i\}$. Then

$$||x||^{\ell} e^{b\cdot x} \leq K'_{\ell,\varepsilon}\, e^{b\cdot x} \sum_{i=1}^{k} (e^{\varepsilon e_i\cdot x} + e^{-\varepsilon e_i\cdot x}) \leq K'_{\ell,\varepsilon} \sum_{i=1}^{k} (e^{(b+\varepsilon e_i)\cdot x}$$

$$+ e^{(b-\varepsilon e_i)\cdot x}) \leq 2k\ K'_{\ell,\varepsilon} \max(e^{b_i\cdot x}) \leq 2k\ K'_{\ell,\varepsilon} \sum_{i=1}^{I} e^{b_i\cdot x} \quad .$$

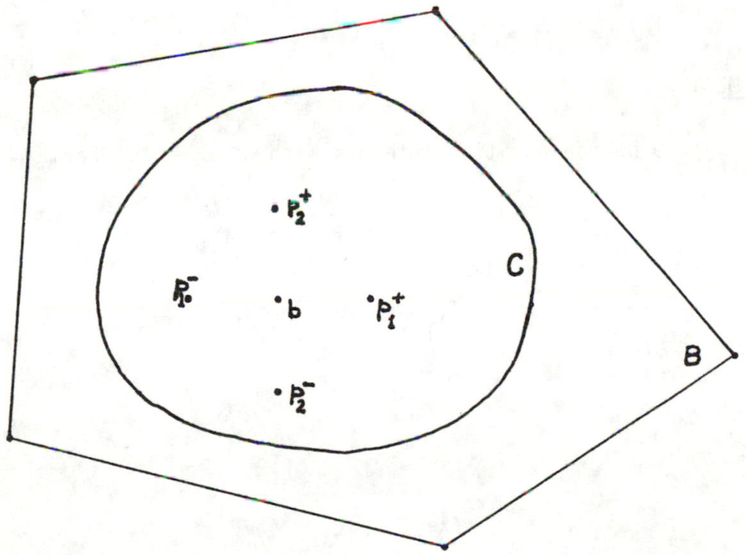

Figure 2.1(1): B, C, and $p_{i\pm} = b \pm \varepsilon e_i$ for the proof of Lemma 2.1.

This proves (1), with $K_\ell = 2k\, K'_{\ell,\varepsilon}$.

Note that (2) may also be written

$$\left| \frac{e^{(b-b_0)\cdot x} - 1}{b - b_0} \right| \le K_1 \sum_{i=1}^{I} e^{(b_i - b_0)\cdot x} \quad .$$

Hence it suffices to prove (2) in the case where $b_0 = 0$, (so that $e^{b_0 \cdot x} = 1$) and we make this assumption below. Note that $re^r - e^r + 1 \ge 0$ and also that for $r \le 0$ $1-e^r < |r|$. Using the first inequality when $b\cdot x > 0$ and the second when $b\cdot x < 0$ yields

$$\left| \frac{e^{b\cdot x} - 1}{||b||\ ||x||} \right| \le \left| \frac{e^{b\cdot x} - 1}{b\cdot x} \right|$$

$$\le \frac{\max(b\cdot x e^{b\cdot x}, |b\cdot x|)}{|b\cdot x|} .$$

Hence

$$\left| \frac{e^{b\cdot x} - 1}{||b||} \right| \le ||x|| e^{b\cdot x} + ||x|| \le 2K_1 \sum_{i=1}^{p} e^{b_i \cdot x}$$

by (1) since $b \in C$ and $0 \in C$. ||

FORMULAS FOR MOMENTS

Let $\ell_i \ge 0$ be non-negative integers with $\sum_{i=1}^{k} \ell_i = \ell$. Formal calculation yields

(1) $$\frac{\partial^\ell}{\prod_{i=1}^{k} \partial\theta_i^{\ell_i}} \lambda(\theta) = \int \left(\prod_{i=1}^{k} x_i^{\ell_i}\right) e^{\theta\cdot x} \nu(dx) \quad .$$

In particular

(2) $$\nabla\lambda(\theta) = \int x e^{\theta\cdot x} \nu(dx) \quad .$$

These calculations are justified by the following theorem.

2.2 Theorem

Suppose $\theta_0 \in N^\circ$. Then all derivatives of λ and of ψ exist at

θ_0. They are given by the above expressions (1), (2) derived by formally differentiating under the integral sign.

Proof. We prove only (2). (The proof of the general formula (1) is similar and proceeds by induction on ℓ. See Exercise 2.2.1.) Let $\theta_0 \in N^\circ$. Then there is a $B = \text{conhull}\{\theta_i : i=1,\ldots,I\} \subset N^\circ$ and $C \subset B^\circ$, C compact, with $\theta_0 \in C^\circ$.

Let

$$d(\theta, x) = \frac{e^{\theta\cdot x} - e^{\theta_0\cdot x} - (\theta-\theta_0)\cdot xe^{\theta_0\cdot x}}{||\theta - \theta_0||} . \tag{3}$$

By Lemma 2.1

$$\sup_{\theta\in C} |d(\theta, x)| \le 2K_1 \Sigma e^{\theta_i\cdot x} . \tag{4}$$

Also

$$|d(\theta, x)| \to 0 \quad \text{as} \quad \theta \to \theta_0 \tag{5}$$

since $\nabla e^{\theta\cdot x}\big|_{\theta=\theta_0} = xe^{\theta_0\cdot x}$. Hence

$$\int d(\theta, x)\nu(dx) \to 0 \quad \text{as} \quad \theta \to \theta_0$$

by the dominated convergence theorem, so that

$$\frac{\lambda(\theta) - \lambda(\theta_0) - (\theta - \theta_0)\cdot \int xe^{\theta_0\cdot x}\nu(dx)}{||\theta - \theta_0||} \underset{\theta\to\theta_0}{\to} 0 , \tag{6}$$

which proves (1). ||

Theorem 2.2 immediately yields the following fundamental formulae. For $f : R^k \to R$ introduce the notation D_2f for the k×k matrix $(\frac{\partial^2 f}{\partial x_i \partial x_j})$. An alternate expression is $\nabla'\nabla f$ since ∇' converts each element of the (column) vector $\frac{\partial f}{\partial x_i}$ into the row vector $(\partial(\frac{\partial f}{\partial x_i})/\partial x_j : j=1,\ldots,k)$, and hence $D_2f = \nabla'\nabla f$.

2.3 Corollary

Consider a standard exponential family. Let $\theta \in N^\circ$. Then

(1) $$E_\theta(X) = \nabla\psi(\theta)$$

(2) $$\operatorname{cov}_\theta X = D_2\psi(\theta) = \nabla'\nabla\psi(\theta) \quad .$$

Notation. In the sequel we frequently use the notation

(1') $$\xi(\theta) = \nabla\psi(\theta) = E_\theta(X)$$

and

(2') $$\Sigma(\theta) = D_2\psi(\theta) = \Sigma_\theta(X) \; .$$

Proof. Calculating formally,

$$\nabla\psi(\theta) = \int x e^{\theta\cdot x}\nu(dx)/\int e^{\theta\cdot x}\nu(dx)$$

$$= E_\theta(X) \; .$$

The calculation is justified by Theorem 2.2. This proves (2). The proof of (1) is similar. ||

2.4 Examples

The reader is invited to use Corollary 2.3 to calculate the familiar formulae for mean and variance in the classic exponential families such as (univariate) normal, multinomial, Poisson, gamma, negative binomial, etc..

For the multivariate normal distribution Corollary 2.3 provides a benefit in the reverse direction. Let Y be m-variate normal (μ, Σ), as in Example 1.14. Fix $\mu = 0$. Direct calculation (not using Corollary 2.3) yields the familiar result

(1) $$E(Y_iY_j) = \sigma_{ij} = (-(\mathcal{Q})^{-1})_{ij} = -\theta^{ij}$$

when $\mu = 0$, where $\mathcal{Q}^{-1} = (\theta^{ij})$. Calculation using Corollary 2.3 and the formula 1.14(3) for the cumulant generating function thus yields for $i \neq j$

(2) $$\frac{\partial}{\partial\theta_{ij}} (-\tfrac{1}{2})\log|-\mathcal{Q}| = -\theta^{ij}/(1 + \delta_{ij})$$

since the corresponding canonical statistics are $Y_iY_j/(1+\delta_{ij})$. Let $B = -\mathcal{Q}$. Then (2) shows that for any positive definite symmetric matrix, B,

(3) $$\frac{\partial}{\partial b_{ij}} \log|B| = 2b^{ij}/(1+\delta_{ij}) \quad \text{where} \quad B^{-1} = (b^{ij}) .$$

Hence, also,

(4) $$\frac{\partial}{\partial b_{ij}} |B| = 2b^{ij}|B|/(1+\delta_{ij}) . \qquad ||$$

The convexity of ψ together with Theorem 2.2 yields the following useful result.

2.5 Corollary

Let $\theta_1,\theta_2 \in N^\circ$. Then

(1) $$(\theta_1 - \theta_2)\cdot(\xi(\theta_1) - \xi(\theta_2)) \geq 0 .$$

Equality holds in (1) if and only if $P_{\theta_1} = P_{\theta_2}$. Consequently $\xi(\theta_1) = \xi(\theta_2)$ if and only if $P_{\theta_1} = P_{\theta_2}$. (If $\{p_\theta\}$ is minimal this happens only when $\theta_1 = \theta_2$.)

Proof. ψ is convex. Hence the directional derivative of ψ in direction $\theta_1 - \theta_2$ is non-decreasing as one moves along the line from θ_2 to θ_1. That is,

(2) $$(\theta_1 - \theta_2)\cdot\nabla\psi(\theta_2 + \rho(\theta_1 - \theta_2)) = (\theta_1 - \theta_2)\cdot\xi(\theta_2 + \rho(\theta_1 - \theta_2))$$

is non-decreasing in ρ. This yields (1).

If $P_{\theta_1} \neq P_{\theta_2}$ then ψ is strictly convex on the line joining θ_2 and θ_1. Hence (2) is strictly increasing for $\rho \in (0,1)$. This yields the remaining assertions of the corollary. (The parenthetical assertion is contained in Theorem 1.13.) ||

The final corollary to Theorem 2.2 establishes the possibility of differentiating inside the integral sign for expectations involving exponential families. The result is stated only for real valued statistics, but obviously generalizes to higher dimensional statistics.

2.6 Corollary

Let $T : R^k \to R$. Let

(1) $$N(T) = \{\theta : \int |T(x)| e^{\theta \cdot x} \nu(dx) < \infty\} .$$

Then $N(T)$ is convex. Define

(2) $$h(\theta) = \int T(x) e^{\theta \cdot x} \nu(dx) = e^{\psi(\theta)} E_\theta(T(X))$$

for $\theta \in N(T)$. Then all derivatives of h exist at every $\theta \in N^\circ(T)$, and they may be computed under the integral sign. In particular

(3) $$\nabla E_\theta(T(X)) = \int (x - \xi(\theta)) T(x) \exp(\theta \cdot x - \psi(\theta)) \nu(dx) .$$

Proof. Suppose $T(x) \geq 0$. Applying Theorem 2.2 to the measure $\omega(dx) = T(x)\nu(dx)$ yields the desired results. For general T the corollary follows upon using the above to separately treat T^+ and T^-. ||

Note that if T and $|T|^{-1}$ are bounded then $N(T) \supset N$.

ANALYTICITY

The moment generating function is analytic. This fact is implicit in the proof of Theorem 2.2. As a preliminary we extend the definition of λ and ψ to the complex domain.

Let

$$\lambda : \mathbb{C}^k \to \mathbb{C}$$

be defined by the same expression as previously, i.e.

(1) $$\lambda(\theta) = \int \exp(\theta \cdot x) \nu(dx) .$$

For $\theta \in \mathbb{C}^k$ let Re θ denote the vector with coordinates (Re $\theta_1, \ldots,$ Re θ_k). Note that for $x \in R^k$

(2) $$|e^{\theta \cdot x}| = e^{(\mathrm{Re}\,\theta) \cdot x} .$$

Hence $\lambda(\theta)$ exists for $\operatorname{Re}\theta \in N$.

2.7 Theorem

$\lambda(\theta)$ is analytic on $\{\theta \in \mathbb{C}^k : \operatorname{Re}\theta \in N^\circ\}$.

Proof. Lemma 2.1 (and its proof) apply for $b \in \mathbb{C}^k$, $x \in R^k$. Similarly the proof of Theorem 2.2(2) is valid verbatim for $\theta \in \mathbb{C}^k$. Thus $\nabla\lambda(\theta)$ exists for $\operatorname{Re}\theta \in N^\circ$ (and has the expression 2.2(2)). This implies that λ is analytic on this domain. ||

Two important properties of analytic functions are: (i) they can be expanded in a Taylor series; and (ii) they are analytic in each variable separately. Thus, for a fixed value of $(\theta_2,\ldots,\theta_k)$, $\lambda((\cdot,\theta_2,\ldots,\theta_k))$ is analytic. $\lambda((\cdot,\theta_2,\ldots,\theta_k))$ is determined by its values on any subset having an accumulation point. This is the basis for the following result.

2.8 Lemma

Let $T : R^k \to R$, and let

(1) $\quad h(\theta) = \int T(x)e^{\theta\cdot x}\nu(dx), \quad$ for $\operatorname{Re}\theta \in N(T)$, as defined in 2.6(1).

Then h is analytic on $\{\theta \in \mathbb{C}^k : \operatorname{Re}\theta \in N^\circ(T)\}$.

Let L be a line in R^k, and let $B \subset L \cap N(T)$ be any subset of $L \cap N(T)$ having an accumulation point in $N^\circ(T)$. Then

(2) $$h(\theta) = 0 \quad \forall\, \theta \in B$$

implies $h(\theta) = 0$ for all $\theta \in R^k$ such that $\theta \in L \cap N^\circ(T)$.

Proof. The first assertion follows upon applying Theorem 2.7 to $T^+(x)\nu(dx)$, and $T^-(x)\nu(dx)$.

Next, one may apply linked affine transformations as in Proposition 1.6. Because of this it suffices to consider the case where $L = \{\theta \in R^k : \theta_2 = \ldots = \theta_k = 0\}$. $h((\theta_1,0,\ldots,0))$ is an analytic function of $\theta_1 \in \mathbb{C}$, as already noted. Hence (2) implies $h(\theta,0,..,0) \equiv 0$ on its domain of analyticity, which is $\{(\theta,0,\ldots,0) : \operatorname{Re}\theta \in L \cap N^\circ(T)\}$. This proves the analyticity, which is $\{(\theta,0,\ldots,0): \text{Re } \theta \in L \cap N^o(T)\}$. This proves the second assertion. ||

Note that, more generally, if B is as above then the values of h on B uniquely determine by analytic continuation its value on all of $L \cap N^o(T)$.

(Straight lines play a special role in the above lemma. However we note that there is a valid generalization of the above lemma in which L can be replaced by a suitable one dimensional curve determined as the locus of points satisfying (n - 1) simultaneous analytic equations (C. Earle (1980), personal communication). For example L may be taken to be the curve $x_1^2 + x_2^2 = 1,\ x_3 = \ldots = x_k = 0.$)

2.9 Example

A question which arises, in statistical estimation theory, is whether the positive part James-Stein estimator for an unknown normal mean,

$$\delta(x) = (1 - (k-2)||x||^{-2})^+ x, \quad x \in R^k,$$

can possibly be generalized Bayes for squared error loss. This is equivalent to asking whether $\delta(\cdot)$ can be the gradient of a cumulant generating function for some measure $\nu(d\theta)$ having $N = R^k$. (Note interchange of roles of θ and x.) See Theorem 4.16. The answer is, "No." To see this note that $\delta(x) \equiv 0$ for $||x|| \leq 1$. Hence if $\delta(x) = \nabla\psi(x) = \nabla\lambda(x)/\lambda(x)$ for $||x|| < 1$ it follows by analyticity that $\psi(x) \equiv 0$ on its domain of analyticity, which in this case is R^k. This implies $\delta(x) \equiv 0$, a contradiction. ||

2.10 Example

The question arises in the theory of hypothesis tests as to whether the unit square,

$$S = \{x \in R^k : |x_i| \leq 1\}, \quad k \geq 2,$$

can be a Bayes acceptance region for testing the mean of a normal distribution. Placed in a general context, the question is whether there exist two distinct non-zero finite measures G_0 and G_1 (concentrated on disjoint sets Θ_0 and $\Theta_1 \subset R^k$) such that

(1) $$d(x) = \int e^{\theta\cdot x-\theta^2/2}(G_0(d\theta) - G_1(d\theta)) \geq 0 \quad \text{if } x \in S,$$

and $d(x) \leq 0$ if $x \notin S$. The answer is, "No."

Proof. Let $\mu_i(d\theta) = e^{-\theta^2/2}G_i(d\theta)$, $i = 0,1$. Then $d(x) = \lambda_0(x) - \lambda_1(x)$ where λ_i is the moment generating function of μ_i. Note that $N_{\mu_i} = R^k$, $i = 0,1$. Hence $d(\cdot)$ is analytic on R^k.

For convenience consider only the case $k = 2$. Expand d in a Taylor series about (1,1) as

$$d((1, 1) + (y_1, y_2)) = \sum_{i=0}^{\infty} \sum_{j=0}^{i} a_{j,i-j}\, y_1^j y_2^{i-2}$$

($a_{00} = 0$ since $d((1, 1)) = 0$.) Let i' be the smallest index for which $\sum_{j=0}^{i'} |a_{j,i'-j}| > 0$. i' exists since $d \not\equiv 0$ if (1) is valid.

Suppose i' is even. Then for $y_i \geq 0$, $i = 1,2$,

(2) $$\sum_{j=0}^{i'} a_{j,i'-j}\, y_1^j y_2^{i'-j} = \sum_{j=0}^{i'} a_{j,i'-j}(-y_1)^j(-y_2)^{i'-j}.$$

There are values (y_1, y_2) in the first quadrant for which (2) $\neq 0$, since (2) is a non-zero homogeneous polynomial. Suppose (y_1^0, y_2^0) is such a value. Then

$$|\rho|^{-i'} d((1, 1)) + (\rho y_1^0, \rho y_2^0)) = \sum_{j=0}^{i'} a_{i',i'-j}(y_1^0)^j(y_2^0)^{i'-j} + o(\rho)$$
$$= c + o(\rho) \quad \text{as } |\rho| \to 0$$

with $c \neq 0$. If $c > 0$ it follows that $d((1, 1)) + (\rho y_1^0, \rho y_2^0)) > 0$ for $\rho > 0$ sufficiently small; and this would contradict (1). If $c < 0$ it follows that $d((1, 1) + (\rho y_1^0, \rho y_2^0)) < 0$ for $\rho < 0$ sufficiently small; and this would also contradict (1).

If i' is odd analogous reasoning yields

$$||y||^{-i'} d((1, 1) + (y_1, -y_2)) = ||y||^{-i'} d((1, 1) + (-y_1, y_2)) + o(1)$$

as $||y|| \to 0$, and that there are values of $(y_1^0, -y_2^0) > 0$ for which $\lim_{\rho\downarrow 0} ||y||^{-i'} d((1, 1) + \rho(y_1^0, -y_2^0)) \neq 0$. It follows that there are values of y in either the fourth quadrant or the second quadrant for which $d((1, 1) + y) > 0$. This again contradicts (1).

Hence (1) is impossible. ||

COMPLETENESS

2.11 Remarks

A family $\{F_\theta : \theta \in \Theta\}$ of probability distributions (or their associated densities, if these exist) is called statistically *complete* if $T : R^k \to R$ with

$$(1) \qquad \int T(x)F_\theta(dx) = 0 \qquad \forall\, \theta \in \Theta$$

implies

$$(2) \qquad T(x) = 0 \quad \text{a.e.} \quad (F_\theta) \qquad \forall\, \theta \in \Theta \quad .$$

(Implicit in (1) is the condition that $\int |T(x)|F_\theta(dx) < \infty \quad \forall\, \theta \in \Theta$.)

Standard exponential families are complete if the parameter space is large enough. This result, which is equivalent to the uniqueness theorem for Laplace transforms, is proved in Theorem 2.12. (The uniqueness theorem for Laplace transforms states that if $N_\mu^\circ \cap N_\nu^\circ \neq \phi$ then $\lambda_\mu = \lambda_\nu$ if and only if $\mu = \nu$.) The most convenient way to prove this theorem seems to be to invoke the uniqueness theorem for Fourier-Stieltjes transforms (equals characteristic functions) which is described in the next paragraph.

Let $Im = \{bi \in \mathbb{C} : b \in R\}$ denote the pure imaginary numbers. Let F be a finite (non-negative) measure on R^k. The function $\kappa : R^k \to \mathbb{C}$ defined by

$$\kappa_F(b) = \lambda_F(bi) \qquad b \in R^k$$

is the *Fourier-Stieltjes transform* (or, Fourier transform, or, *characteristic function*) of F. Hence λ_F restricted to the domain $(Im)^k$ is equivalent to κ_F. Note that κ_F always exists (i.e. $Re((Im)^k) = 0 \subset N$). The uniqueness theorem for Fourier transforms is as follows.

Theorem. Let F and G be two finite non-negative measures on R^k. Then F = G if and only if $\kappa_F \equiv \kappa_G$ (i.e. $\lambda_F(bi) = \lambda_G(bi)\ \forall\, b \in R^k$).

Proof. This is a standard result in the theory of characteristic functions. Proofs abound. A quick proof may be found in Feller (1966, XV,3). (This proof is explicitly for R, but generalizes immediately to R^k.) ||

Here is the classic result on completeness of exponential families.

2.12 Theorem

Let $\{p_\theta\colon \theta \in \Theta\}$ be a standard exponential family. Suppose $\Theta^\circ \neq \phi$. Then $\{p_\theta\}$ is complete.

Proof. Let $\theta_0 \in \Theta^\circ$. One may translate coordinates using Proposition 1.6 so that $\theta_0 = 0$. There is thus no loss of generality in assuming $\theta_0 = 0$.

Suppose $\int T(x)p_\theta(x)\nu(dx) = 0\ \forall\, \theta \in \Theta$. Then, letting $T = T^+ - T^-$,

(1) $$\int T^+(x)e^{\theta\cdot x}\nu(dx) = \int T^-(x)e^{\theta\cdot x}\nu(dx) \qquad \forall\ \theta \in \Theta .$$

Let $F(dx) = T^+(x)\nu(dx)$, $G(dx) = T^-(x)\nu(dx)$. Then (1) becomes

(2) $$\lambda_F(\theta) = \int e^{\theta\cdot x}F(dx) = \int e^{\theta\cdot x}G(dx) = \lambda_G(\theta) \qquad \forall\ \theta \in \Theta .$$

Both $\lambda_F(\cdot)$ and $\lambda_G(\cdot)$ are analytic on the domain $\Theta^\circ \times (\mathrm{Im})^k$. (2) states that they agree on $\Theta \times 0 \subset \Theta \times (\mathrm{Im})^k$. Hence $\lambda_F(x) = \lambda_G(z)$ for all z such that re $z \in \Theta^\circ$. (This follows directly from analyticity. Alternately one may apply the second half of Lemma 2.8 to all lines which intersect Θ .) In particular

(3) $$\lambda_F(0 + bi) = \lambda_G(0 + bi) \qquad \forall\, b \in R^k$$

since $0 \in \Theta^\circ$. Thus, F = G by Theorem 2.11. This says that $T^+(x)\nu(dx) = T^-(x)\nu(dx)$, which implies $T^+ = T^-$ a.e.(ν), which implies $T = 0$ a.e.(ν). Hence $\{p_\theta\}$ is complete. ||

Note from the above that any canonical family is complete. From this we derive:

2.13 Corollary

A standard family with $N^\circ \neq \phi$ is uniquely determined by its Laplace transform (or by its cumulant generating function).

Note that the corollary applies to all minimal families since they always have $N^\circ \neq \phi$.

Proof. Consider the standard families in R^k generated by the measures μ and ν. Suppose $N^\circ_\mu \neq \phi$ and $\psi_\mu = \psi_\nu$. Then $N_\mu = N_\nu = N$.

Let $\omega = (\mu + \nu)/2$. Then, ω generates an exponential family with $\lambda_\omega = (\lambda_\mu + \lambda_\nu)/2$. Hence $N_\omega = N$ and $\psi_\omega = \psi_\mu = \psi_\nu$.

Let $T = \frac{d\mu}{d\omega} - \frac{d\nu}{d\omega}$. Then

$$\int T(x)e^{\theta\cdot x-\psi_\omega(\theta)}\omega(dx) = (\tfrac{1}{2})(\int e^{\theta\cdot x-\psi_\mu(\theta)}\mu(dx) - \int e^{\theta\cdot x-\psi_\nu(\theta)}\nu(dx))$$

$$= \frac{1}{2} - \frac{1}{2} = 0 \qquad \forall\ \theta \in N$$

Hence $T = 0$ a.e.(ω) by Theorem 2.12; which implies $\mu = \nu$. ||

Theorem 2.12 has many other important applications in statistics. It plays an important role, for example, in the theory of unbiased estimates and in the construction of unbiased tests. Some aspects of this role are described in the exercises and in succeeding chapters.

MUTUAL INDEPENDENCE

Lehamnn (1959, p. 162-163) describes a nice proof of the independence of $\bar{X}$, S^2 in a normal sample. A different but related proof is a special instance of an argument which applies in several important exponential families. (See Example 2.15.) The basic parts of the argument are due to Neyman (1938) and Basu (1955), but the full result in Theorem 2.14, below, was only recently proved by Bar-Lev (1983) and by Barndorff-Nielsen and Blaesild (1983). The proof below follows that in the second of these papers. See the exercises for an additional related result of Bar-Lev and for several applications of this theorem.

Through most of this subsection we consider the situation where

θ and x are, respectively, partitioned as $\theta' = (\theta'_{(1)}, \theta'_{(2)})$, $x' = (x'_{(1)}, x'_{(2)})$. As in Sections 1.7 and 1.15, problems not in this form can sometimes be reduced to this form through use of linked linear transformations on θ and x. Where convenient, we write $\psi(\theta) = \psi(\theta_{(1)}, \theta_{(2)})$. We use the notation $Y \sim \text{Expf}\ (\theta)$ to mean that the distributions of Y form a standard exponential family with natural parameter θ. We also use the notation $X \perp Y$ to mean that X and Y are independent.

2.14 Theorem

Let $X \sim \text{Expf}\ (\theta)$ with $\theta^\circ \in \Theta^\circ$. Let $X' = (X'_{(1)}, X'_{(2)})$ where $X_{(i)}$ is k_i dimensional, and let $h(X_{(1)})$ be a k_2 dimensional statistic. Let

$$\rho_1(\theta_{(1)}, \theta_{(2)}) = \log E_\theta\ (\exp((\theta_{(1)} - \theta^\circ_{(1)}) \cdot X_{(1)}$$

$$+ (\theta_{(2)} - \theta^\circ_{(2)}) \cdot h(X_{(1)}))) \tag{1}$$

$$\rho_2(\theta_{(2)}) = \log E_\theta\ (\exp((\theta_{(2)} - \theta^\circ_{(2)}) \cdot (X_{(2)} - h(X_{(1)}))) .$$

Then the following conditions are equivalent:

$$X_{(1)} \perp (X_{(2)} - h(X_{(1)})) \quad \text{under} \quad \theta^\circ \tag{2}$$

$$X_{(1)} \perp (X_{(2)} - h(X_{(1)})) \quad \text{for all} \quad \theta \in \Theta \tag{2'}$$

$$\psi(\theta_{(1)}, \theta_{(2)}) = \rho_1(\theta_{(1)}, \theta_{(2)}) + \rho_2(\theta_{(2)}) \quad \forall \quad \theta \in \Theta \tag{3}$$

$$(X_{(2)} - h(X_{(1)})) \sim \text{Expf}\ (\theta_{(2)}) \tag{4}$$

$$(X_{(1)}, h(X_{(1)})) \sim \text{Expf}\ (\theta_{(1)}, \theta_{(2)}) \quad . \tag{5}$$

Proof. For convenience, assume without loss of generality that $\theta^\circ = 0$. (See Proposition 1.6.) Let ω denote the joint distribution under 0 of $V = (X_{(1)}, h(X_{(1)}), X_{(2)} - h(X_{(1)}))$. Consider the standard exponential family generated by ω, with natural parameter space N_V. Note that, in general,

$\{X_{(1)} \perp (X_{(2)} - h(X_{(1)}))\} \Leftrightarrow \{(X_{(1)}, h(X_{(1)})) \perp (X_{(2)} - h(X_{(1)}))\}$. The equivalence of (2) and (2') is seen in this fashion to be a special case of Exercise 1.7.1.

(2) $\Rightarrow$ (3) follows from a direct calculation.

(3) $\Rightarrow$ (2): Let ω_1 denote the distribution under $\theta^\circ = 0$ of $(V_{(1)}, V_{(2)}) = (X_{(1)}, h(X_{(1)}))$ and ω_2 that of $V_{(3)} = X_{(1)} - h(X_{(1)})$. Let $\omega^* = \omega_1 \times \omega_2$. Then the cumulant generating function ψ^* of ω^* satisfies

$$\psi^*(\theta_{(1)}, \theta_{(2)}, \theta_{(2)}) = \rho_1(\theta_{(1)}, \theta_{(2)}) + \rho_2(\theta_{(2)}), \quad (\theta_{(1)}, \theta_{(2)}) \in \Theta \quad .$$

Furthermore, the cumulant generating function of the linear function $(V_{(1)}, V_{(2)} + V_{(3)})$ is ψ^{**} given by

$$\psi^{**}(\theta_{(1)}, \theta_{(2)}) = \psi^*(\theta_{(1)}, \theta_{(2)}, \theta_{(2)}) = \rho_1(\theta_{(1)}, \theta_{(2)}) + \rho_2(\theta_{(2)})$$

$$= \psi(\theta_{(1)}, \theta_{(2)}) , \quad \theta \in \Theta \quad .$$

It follows from Corollary 2.13, since $\Theta^\circ \neq \phi$, that $(V_{(1)}, V_{(2)} + V_{(3)})$ has the same distribution under θ° as $(X_{(1)}, X_{(2)})$. Thus $(X_{(1)}, X_{(2)} - h(X_{(1)}))$ has the same joint distribution under θ° as $(V_{(1)}, V_{(2)} + V_{(3)} - h(V_{(1)}))$. But, $V_{(2)} + V_{(3)} - h(V_{(1)}) = V_{(3)}$. Hence $X_{(1)} \perp (X_{(2)} - h(X_{(1)}))$ under θ° since $V_{(1)} \perp V_{(3)}$.

(2) $\Rightarrow$ (4) and (5), as can be seen by direct calculation of the marginal distributions involved via the standard formulae (6) and (8), below.

(4) $\Rightarrow$ (2): The marginal density of $V_{(3)} = X_{(2)} - h(X_{(1)})$ relative to the marginal distribution ω_2 is

(6) $$q_\theta(v_{(3)}) = \int \exp(\theta_{(1)} \cdot v_{(1)} + \theta_{(2)} \cdot h(v_{(1)})$$
$$+ \theta_{(2)} \cdot v_{(3)} - \psi(\theta))\, \omega(dv_{(1)} \mid v_{(3)}) \quad \text{(a.e.)}$$

where $\omega(\cdot|\cdot)$ denotes the indicated conditional distribution. By (4) $q_\theta(v_{(3)}) = \exp(\theta_{(2)} v_{(3)} - \rho_2(\theta_{(2)}))$ (a.e.). Setting $\theta_{(2)} = 0$ yields

(7) $\exp(\psi(\theta_{(1)}, 0) - \rho_2(0))$

$$= \int \exp(\theta_{(1)} \cdot v_{(1)})\, \omega(dv_{(1)}|v_{(3)}) , \quad (\theta_{(1)}, 0) \in \Theta , \quad \text{(a.e.)} .$$

Here the Laplace transform of $\omega(\cdot|v_{(3)})$ exists on an open set and is independent of $v_{(3)}$ (a.e.). It follows from another application of Corollary 2.13 that $\omega(\cdot|v_{(3)})$ is independent of $v_{(3)}$ (a.e.). So, $V_{(1)}$ is independent of $V_{(3)}$. This verifies (2).

The proof that (5) ⇒ (2) is similar. The marginal joint density of $V_{(1)}, V_{(2)}$ is

(8) $$q'_\theta(v_{(1)}, v_{(2)}) = \int \exp(\theta_{(1)} \cdot v_{(1)} + \theta_{(2)} \cdot h(v_{(1)}) + \theta_{(2)} \cdot v_{(3)}$$

$$- \psi(\theta))\, \omega'(dv_{(3)}|v_{(1)}) \quad \text{(a.e.)} .$$

Setting $\theta_{(1)} = 0$ and cancelling terms in (5) implies

$$\exp(\psi(0, \theta_{(2)}) - \rho(0, \theta_{(2)})) = \int \exp(\theta_{(2)} \cdot v_{(3)})\, \omega'(dv_{(3)}|v_{(1)}) \quad \text{(a.e.)}.$$

Hence, as before $\omega'(\cdot|v_{(1)})$ is independent of $v_{(1)}$ (a.e.), which yields (2). ||

2.15 Examples

(i) Let $Y_1,\ldots,Y_n$ be independent $N(\mu,\sigma^2)$ variables. Then (Example 1.12) $(Y_i, Y_i^2) \sim \text{Expf}(\mu/\sigma^2, -1/2\sigma^2)$. Hence $(\Sigma Y_i, \Sigma Y_i^2) \sim \text{Expf}(\mu/\sigma^2, -1/2\sigma^2)$. Also $(\Sigma Y_i, (\Sigma Y_i)^2/n) \sim \text{Expf}(\mu/\sigma^2, -1/2\sigma^2)$. This verifies 2.14(5). Hence $\Sigma Y_i - \Sigma Y_i^2/n = \Sigma(Y_i - \bar{Y})^2 \sim \text{Expf}(-1/2\sigma^2)$ and is independent of T_n by 2.14(4) and 2.14(2').

(ii) Similarly, let $X_1,\ldots,X_n$ be independent $\Gamma(\alpha, \sigma)$. Then (Example 1.12) $\Gamma(\Sigma X_i, \Sigma \ln X_i) \sim \text{Expf}(-1/\sigma, n\alpha)$. The marginal distribution of X_i is also $(n\alpha, \sigma)$; hence $(\Sigma X_i, \ln \Sigma X_i) \sim \text{Expf}(-1/\sigma, n\alpha)$. Again, Theorem 2.14 yields that $(\Sigma \ln X_i - \ln \Sigma X_i) \perp \Sigma X_i$. This is often re-expressed in the form $\tilde{X}/\bar{X} \perp \bar{X}$ where here $\tilde{X} = (\prod_{i=1}^{n} X_i)^{1/n}$ denotes the geometric mean of the observations. Also, $\ln(\tilde{X}/\bar{X}) \sim \text{Expf}(n\alpha)$. See the Exercises for a double extension of this conclusion.

There are further applications of this theorem. For some of these see the exercises and the references cited above. In particular there are several applications to problems involving the inverse Gaussian distribution. See Chapter 3.

CONTINUITY THEOREM

The continuity theorem for Laplace transforms refers to the limiting behavior of a sequence of measures and the associated Laplace transforms. We first need a standard definition and some related remarks.

2.16 Definition

Consider R^k. Let C denote the space of continuous (real-valued) functions on R^k. Let $C_0 \subset C$ denote the subspace of continuous functions with compact support -- i.e.

$$c(x) = 0 \quad \text{for} \quad ||x|| > r\,, \quad \text{some} \quad r < \infty\,.$$

A (non-negative) measure ν is called *locally finite* if $\nu(\{x : ||x|| \le r\}) < \infty \; \forall \; r \in R$. Except where specifically noted, all measures are assumed to be locally finite, σ-finite, and non-negative. Let $\{\nu_n\}$ be a sequence of measures. We say

$$\nu_n \to \nu \quad \text{(weak*)} \quad \text{if}$$

(1) $$\int c(x)\nu_n(dx) \to \int c(x)\nu(dx) \quad \forall\, c \in C_0\,.$$

Here are several important facts concerning weak* convergence.

For ν finite let V_ν denote the *cumulative distribution function*: $V_\nu(t) = \nu(\{x : x_i \le t_i, \; i=1,\ldots,k\})$.

(i) Then $\nu_n \to \nu$ if and only if

(2) $$V_{\nu_n}(t) \to V_\nu(t) \quad \forall\, t \in R^k \quad \text{at which } V_\nu(\cdot) \text{ is continuous.}$$

(ii) Suppose $\nu_n \to \nu$. Then $\liminf_{n\to\infty} \nu_n(R^k) \ge \nu(R^k)$. Suppose there is a $c \in C$, $c \geq 0$, with

(3) $$\lim_{||x|| \to \infty} c(x) = \infty$$

such that

(3') $$\limsup_{n \to \infty} \int c(x)\nu_n(dx) < \infty \quad .$$

Then

(4) $$\lim_{n \to \infty} \nu_n(R^k) = \nu(R^k) < \infty \quad .$$

(iii) Furthermore, (4) implies

(5) $$\int c(x)\nu_n(dx) \to \int c(x)\nu(dx)$$

for all bounded $c \in C$. (Condition (3), (3') is sometimes referred to by saying the sequence is *tight*.)

(iv) If $\nu_n \geq 0$ is any bounded sequence (i.e. $\limsup_{n \to \infty} \nu_n(R^k) < \infty$) then there is a subsequence $\{\nu_{n_i}\}$ and a finite measure ν such that $\nu_{n_i} \to \nu$.

For a proof of these facts see Neveu (1965).

2.17 Theorem

Let $S \subset R^k$ and let $B = \text{conhull } S$. Suppose $B^\circ \neq \phi$. Let ν_n be a sequence of measures on R^k such that

(1) $$\liminf_{n \to \infty} \sup_{b \in S} \lambda_{\nu_n}(b) < \infty \quad \forall \ b \in S \ .$$

Let $b_0 \in B^\circ$. Then there exists a subsequence $\{n_i\}$ and a locally finite measure ν such that

(2) $$e^{b_0 \cdot x} \nu_{n_i}(dx) \to e^{b_0 \cdot x}\nu(dx) \quad ,$$

and

(3) $$\lambda_{\nu_{n_i}}(b) \to \lambda_\nu(b) \quad \forall \ b \in B^\circ \ .$$

The convergence in (3) is uniform on compact subsets of B°.

(Condition (3) is of course equivalent to

$$\psi_{\nu_{n_i}}(b) \to \psi_\nu(b) \quad \forall \ b \in B^\circ \qquad .$$

Condition (1) implies the measures ν_n are locally finite.)

Remark. Lemma 2.1 together with (3) shows that

(4) $$\int x e^{b \cdot x} \nu_{n_i}(dx) \to \int x e^{b \cdot x} \nu(dx), \qquad b \in B^\circ \quad ,$$

and similarly for higher moments of x. Hence

(5) $$\nabla\lambda_{\nu_{n_i}}(b) \to \nabla\lambda_\nu(b), \qquad b \in B^\circ$$

and similarly for higher order partial derivatives of λ. See Exercise 2.17.1.

Similar reasoning also shows that

(6) $$e^{\theta \cdot b}\nu_{n_i}(d\theta) \to e^{\theta \cdot b}\nu(d\theta) \quad \text{weak*} \quad \forall \ b \in B_0 \quad .$$

Hence the measure ν in (2) does not depend on the choice of $b_0 \in B_0$.

Proof. We exploit Proposition 1.6 and assume without loss of generality that $b_0 = 0 \in B^\circ$. It also suffices to assume that B is a convex polytope (i.e. $B = \text{conhull}\ \{b_i : i=1,\ldots,m\}$) since the interior of any convex set is a countable union of such polytopes, and a compact subset of the interior will be contained in one of them.

Now,

$$\lim_{||x|| \to \infty} \sum_{i=1}^{m} e^{b_i \cdot x} = \infty$$

by Lemma 2.1. Thus, for some subsequence $\{n'_j\}$

(7) $$\limsup_{n \to \infty} \int \Big(\sum_{i=1}^{m} e^{b_i \cdot x}\Big)\nu_{n'_j}(dx) < \infty \qquad ,$$

by (1). Hence, the sequence $\{\nu_{n'_j}\}$ is tight, and there exists a further subsequence $\{\nu_{n_i}\}$ and a limiting measure ν such that $\nu_{n_i} \to \nu$. This immediately implies that also $e^{b \cdot x}\nu_{n_i}(dx) \to e^{b \cdot x}\nu(dx)$ for any $b \in R^k$.

Let $b \in B^\circ$. Then $\lim_{||x||\to\infty} (\Sigma e^{b_i \cdot x}/e^{b\cdot x}) = \infty$, again by Lemma 2.1.

As in (7)

$$\limsup_{n\to\infty} \int \frac{\Sigma e^{b_i \cdot x}}{e^{b\cdot x}} e^{b\cdot x} \nu_{n_i}(dx) < \infty.$$

Hence the sequence $e^{b\cdot x}\nu_{n_i}(dx)$ is also tight. This implies $\int e^{b\cdot x}\nu_{n_i}(dx) \to \int e^{b\cdot x}\nu(dx)$, which yields (3).

Let $C \subset B^\circ$ be compact. Then

$$||x||e^{b\cdot x} \le K \sum_{i=1}^{m} e^{b_i \cdot x}$$

by Lemma 2.1. This yields

$$\limsup_{i\to\infty} \sup_{b\in C} ||\nabla\lambda_{\nu_{n_i}}(b)|| \le \limsup_{i\to\infty} K \int \sum_{i=1}^{m} e^{b_i \cdot x} \nu_{n_i}(dx)$$

$$\le \limsup_{i\to\infty} K \sum_{i=1}^{m} \lambda(b_i) < \infty .$$

The functions $\lambda_{\nu_{n_i}}(\cdot)$ are thus uniformly (in $\{n_i\}$) uniformly continuous on C. The convergence in (3) is therefore uniform on C. ||

2.18 Uniform Convergence

Theorem 2.17 shows that if

$$\psi_{\lambda_i}(b) \to \psi_\lambda(b) \quad \text{for all} \quad b \in B^\circ \ne \phi \quad \text{then} \quad \nu_i \to \nu .$$

There is a useful uniform version of this statement. Let

(1) $$\{\nu_{\alpha n} : n=1,\ldots,\alpha \in A\}$$

be a family of sequences of measures and $\{\nu_\alpha : \alpha \in A\}$ be a family of measures. All of these are assumed locally finite. We say

$$\nu_{\alpha n} \to \nu_\alpha \text{ (weak*) uniformly in } \alpha$$

when for each $c \in C_0$

(2) $$\int c(x)\nu_{\alpha n}(dx) \underset{n\to\infty}{\to} \int c(x)\nu_\alpha(dx)$$

uniformly over $\alpha \in A$. For notational convenience in the following, let $V_\alpha = V_{\nu_\alpha}$, etc.

Proposition. Suppose the family of cumulative distribution functions $\{V_\alpha: \alpha \in A\}$ is equicontinuous at every $x \in R^k$. Then $\nu_{\alpha n} \to \nu_\alpha$ uniformly in α if and only if

(3) $$V_{\alpha n} \to V_\alpha \quad \text{uniformly for} \quad \alpha \in A \quad .$$

Proof. The necessity of (3) is proved by applying (2) to continuous functions c satisfying

$$c(x) = \begin{cases} 1 & x_i \le x_{0i} - \delta \quad \text{for all } i=1,\dots,k \\ 0 & x_i > x_{0i} + \delta \quad \text{for some } i=1,\dots,k \end{cases}$$

and then choosing δ sufficiently small.

Conversely, (3) implies $\int g(x)d(V_{\alpha n}(x) - V_\alpha(x)) = \int(V_{\alpha n}(x) - V_\alpha(x))dg(x) \to 0$ uniformly in α for each differentiable $g \in C_0$. If $c \in C_0$ and $\varepsilon > 0$ there is a differentiable $g \in C_0$ with $|g - c| < \varepsilon$. Then $|\int(c(x) - g(x))\, d(V_{\alpha n}(x) - V_\alpha(x))| < 2\varepsilon$ uniformly for all $\alpha \in A$ and all n. Combining these facts yields the uniform convergence of $\nu_{\alpha n}$ to ν_α. ||

Extra care in the proof of the above proposition will show that if the $\{V_\alpha : \alpha \in A\}$ are equicontinuous uniformly over $x \in S$ and $\nu_{\alpha n} \to \nu_\alpha$ uniformly in α then (3) holds uniformly for $\alpha \in A$, $x \in S$.

2.19 Theorem

Let $\{\nu_{\alpha n}\}$ and $\{\nu_\alpha\}$ be as in 2.18(1). Suppose B = conhull S, and $B^\circ \ne \phi$. Let $\lambda_\alpha = \lambda_{\nu_\alpha}$, etc. Suppose

(1) $$\lambda_{\alpha n}(b) \underset{n\to\infty}{\to} \lambda_\alpha(b) \qquad \forall\ b \in S$$

uniformly over $\alpha \in A$, and suppose

(2) $$\sup_{b \in S} \sup_{\alpha} \lambda_\alpha(b) < \infty \quad .$$

Then $\nu_{\alpha n} \to \nu_\alpha$ uniformly over $\alpha \in A$.

Proof. If $\nu_{\alpha n} \not\to \nu_\alpha$ uniformly over $\alpha \in A$, there is a $c \in C_0$ and a sequence α_n such that

(3) $$\lim_{n\to\infty} |\int c(\theta)(\nu_{\alpha_n n}(d\theta)-\nu_{\alpha_n}(d\theta))| > 0 \quad .$$

In view of (3) there exists a subsequence n_i and limiting measures $\nu_1^* \neq \nu_2^*$ such that if we write $\nu_{\alpha_{n_i} n_i} = \omega_i$ and $\nu_{\alpha_{n_i}} = \bar{\omega}_i$ then

(4) $$\omega_i \to \nu_1^* , \qquad \lambda_{\omega_i}(b) \to \lambda_{\nu_1^*}(b) , \qquad b \in B ;$$

and

(5) $$\bar{\omega}_i \to \nu_2^* , \qquad \lambda_{\bar{\omega}_i}(b) \to \lambda_{\nu_2^*}(b) \qquad b \in B \quad ,$$

by Theorem 2.17. (To establish (4) we exploit (2) to guarantee condition 2.17(1) for the sequence $\{\omega_{n_i}\}$.)

Assumption (1) implies $\lambda_{\nu_1^*}(b) = \lambda_{\nu_2^*}(b)$, $b \in B$, which implies $\nu_1^* = \nu_2^*$. This is a contradiction. It follows that $\nu_{\alpha n} \to \nu_\alpha$ uniformly over $\alpha \in A$. ||

TOTAL POSITIVITY

2.20 Definitions

Let $S \subset R$ and $h : S \to R$. Let $\{x_0 < \ldots < x_n\} \subset S$. The sequence $\{x_i \in S : i=0,1,\ldots,n\}$ is called a strictly changing sequence for h having order n if

(1) $$(\operatorname{sgn} h(x_{i-1}))(\operatorname{sgn} h(x_i)) = -1 \qquad i=1,\ldots,n \quad .$$

The number $S^-(h)$ -- *the number of strict sign changes of* h -- is the maximal order of a sequence of strict sign changes of h. Clearly $0 \le S^-(h) \le \infty$. Let $S^-(h) = n < \infty$ and let $\{x_i \in S : i=0,\dots,n\}$ be a strictly changing sequence for h having order n. Then *the (strict) initial sign of* h is

(2) $$IS^-(h) = \operatorname{sgn} h(x_0) .$$

(It is easy to check that this definition is well-formulated -- i.e. does not depend on the chosen strictly changing sequence for h.)

Similarly a sequence $\{x_i \in S : i=0,\dots,n\}$ is called a weakly changing sequence for h having order n if

(3) $$(\operatorname{sgn} h(x_{2i}))(\operatorname{sgn} h(x_{2j+1})) \le 0$$

$$\text{for} \quad i=0,\dots,[n/2], \quad j=0,\dots,[(n-1)/2] .$$

This means that zeros of the sequence $\{\operatorname{sgn} h(x_i) : i=0,1,\dots,n\}$ can be reassigned as either a (+1) or a (-1) in a manner so that the resulting sequence of ±1's alternates in sign. The number $S^+(h)$ is the maximal order of such a sequence. Clearly, $0 \le S^+(h) \le \infty$, and

(4) $$S^+(h) \ge S^-(h) .$$

Let $S^+(h) = n < \infty$ and let $\{x_i \in S : i=0,\dots,n\}$ be a weakly changing sequence for h of order n. Then

(5) $$IS^-(h) = \begin{cases} +1 & \text{if } h(x_{2i}) > 0 \quad \text{for some} \quad i=0,\dots,[n/2] \\ 0 & \text{if } h(x_i) \equiv 0 \quad i=0,\dots,n \\ -1 & \text{if } h(x_{2i}) < 0 \quad \text{for some} \quad i=0,\dots,[n/2] . \end{cases}$$

It can be checked that this definition is well formulated.

<u>2.21 Theorem</u>

Let $\{p_\theta\}$ be a standard one parameter exponential family. Let $g : R \to R$ such that $\nu\{x : g(x) \ne 0\} > 0$. Let

(1) $$h(\theta) = E_\theta(g(x)) , \quad \theta \in N^\circ(g) .$$

Then

(2) $$S^+(h) \le S^-(g) .$$

If equality holds in (2) then

(3) $$IS^+(h) = IS^-(g) .$$

Remark. The domain of h in (1) is restricted to $N^\circ(g)$. The theorem remains true if the domain of h is all of $N(g)$. We leave this generalization as an exercise.

The sign-change-preserving properties (2), (3) are equivalent to "Total Positivity of $\{p_\theta\}$ of order ∞." Karlin (1968) is a very useful, standard reference on this topic. See also Brown, Johnstone, and MacGibbon (1981).

Proof. Let

$$\hat{g}(\theta) = \int e^{\theta x} g(x)dx = e^{\psi(\theta)} h(\theta) .$$

It suffices to prove $\hat{g}$ has the properties of h in (2), (3). The proof is by induction on $n = S^-(g)$. Assume without loss of generality that $IS^-(g) = +1$.

When n = 0 the result is trivial since then $g \ge 0$ and $\nu(\{x : g(x) > 0\}) > 0$ so that $\hat{g}(\theta) > 0$ for all $\theta \in N(h)$, as claimed in (2).

Assume the theorem is true for $n \le N$. Suppose $n = N + 1$. Let $\xi_1 = \inf\{x : g(x) < 0\}$. $\xi_1 > -\infty$ since $IS^-(g) = +1$. Let

$$u(\theta) = \frac{d}{d\theta} (e^{-\theta\xi_1}\hat{g}(\theta)) = \int (x - \xi_1)g(x)e^{\theta x}\nu(dx) .$$

Now, $S^-((x - \xi_1)g(x)) \le N = n - 1$, as can easily be checked from the definition of ξ_1. Hence $S^+(u) \le N$ by the induction hypothesis. Integration yields that $S^+(U) \le N + 1$ where

(4) $$U(\theta) = \int^\theta u(t)dt = e^{-\xi_1\theta} \hat{g}(\theta) .$$

(2) follows from (4). (3) may be verified by concentrating the above argument on the case where $S^+(u) = N$ and $S^+(U) = N + 1$, and using the induction hypothesis

to keep track of $IS^+(u)$ and consequently of $IS^+(U)$. ||

The above property for n = 1 is equivalent to the strict monotone likelihood ratio property. The following is an important consequence of this.

2.22 Corollary

Let $\{p_\theta\}$ be a standard one parameter exponential family. Suppose $g : R \to R$ is non-decreasing and not essentially a constant (ν). Then $E_\theta(g)$ is strictly increasing on $N^\circ(g)$.

(*Remark*. Again, the result is true on the full domain, $N(g)$, but we leave verification of this as an exercise.)

Proof. Let ess inf $g(\cdot) < c <$ ess sup $g(\cdot)$; then $g(\cdot) - c$ satisfies the hypotheses of Theorem 2.21 with $S^-(g-c) = 1$. Hence $E_\theta(g) - c > 0$ (or < 0) for $\theta \in N^\circ(g)$ whenever $\theta > \theta_1(c)$ (whenever $\theta < \theta_1(c)$). It follows that g is strictly increasing on $N^\circ(g)$. ||

It is possible to derive from the above some results concerning sign changes for multidimensional families. In general, these results appear very weak by comparison with their univariate cousins. Here is an example of such a result which will be useful later.

2.23 Corollary

Let $\{p_\theta\}$ be a standard k parameter exponential family. Let $\theta_0 \in N$ and $v \in R^k$. Let $\theta_\rho = \theta_0 + \rho v$. Suppose $g : R^k \to R$ satisfies

$$g(x) \begin{cases} \le 0 & v \cdot x \le \alpha \\ \ge 0 & v \cdot x \ge \alpha \end{cases} \tag{1}$$

for some $\alpha \in R$. Let

$$h(\rho) = E_{\theta_\rho}(g(X)) .$$

Then $S^+(h) \le 1$. If $S^+(h) = 1$ then $IS^+(h) = -1$.

Proof. Apply Theorem 2.22 to the one parameter exponential family $\{p_{\theta_\rho}\}$ of densities of $v \cdot X$. Observe that

$$E_{\theta_\rho}(g(x) \mid v \cdot x = t) = g^*(t)$$

is independent of ρ by Theorem 1.7, and (1) guarantees that $S^-(g^*) \leq 1$. These observations enable the desired application of the theorem. ||

PARTIAL ORDER PROPERTIES

The preceding multidimensional result is not very satisfactory; the hypotheses on h are too restrictive. Better results may be obtained by considering partial orderings and imposing suitable restrictions on the exponential family. We give one simple result as an appetizer for what may be obtained.

For this result define the partial ordering, $\propto$, on R^k by $x \propto y$ if $x_i \leq y_i$, $i=1,\ldots,k$. A function $h : R^k \to R$ is non-decreasing relative to this ordering if $x \propto y$ implies $h(x) \leq h(y)$. The following preparatory lemma is also of independent interest.

2.24 Lemma

Let X have coordinates $X_1,\ldots,X_k$ which are independent random variables with distributions $F_1,\ldots,F_k$, respectively. Suppose h_1, h_2 are non-decreasing relative to the partial ordering $\propto$. Then

$$E(h_1(X)h_2(X)) \geq E(h_1(X))E(h_2(X)) \quad . \tag{1}$$

Proof. The proof is by induction on k. Note that for k = 1 the result is well known. This observation enables one to rewrite and reduce the left side of (1) as

$$\int \cdots \int h_1(x)h_2(x) \prod_{i=1}^{k} F_i(dx_i)$$

$$= \int \cdots \int \left(\int h_1(x)h_2(x)F_k(dx_k)\right) \prod_{i=1}^{k-1} F_i(dx_i)$$

$$\geq \int \cdots \int \left[\int h_1(x)F_k(dx_k)\right]\left[\int h_2(x)F_k(dx_k)\right] \prod_{i=1}^{k-1} F_i(dx_i) \quad .$$

Each function in square brackets is clearly non-decreasing in $(x_1,\ldots,x_{k-1})$. Hence, by induction, (1) is valid. ||

Here is the application to exponential families.

2.25 Theorem

Consider a minimal standard exponential family for which the canonical coordinate variables $X_1,\ldots,X_k$ are independent. Let h be non-decreasing relative to the partial ordering $\propto$. Then $E_\theta(h)$ is a non-decreasing function of θ on $N^\circ(h)$. (This result may be extended to all of $N(h)$.)

Proof. Write

$$\frac{\partial}{\partial\theta_j} E_\theta(h) = \int (x_j - \xi_j(\theta))h(x)p_\theta(x)\nu(dx) \quad .$$

Note that both $x_j - \xi_j(\theta)$ and $h(x)$ are non-decreasing functions of x. Hence

$$\frac{\partial}{\partial\theta_j} E_\theta(h) \geq E\,(X_j - \xi_j(\theta))E_\theta(h(X)) = 0$$

by Lemma 2.24. It follows that $E_\theta(h)$ is non-decreasing in each coordinate of θ and hence (equivalently) is non-decreasing relative to $\propto$. ||

The preceding theorem is merely a sample of the available results. Other assumptions may replace the independence assumption, above. Notably, the conclusion of Lemma 2.24 remains valid if the joint distribution, F, of X has a density f with respect to Lebesgue measure which is monotone likelihood ratio in each pair of coordinates when the others are held fixed. (Exercise.) (There is also a lattice variable version of this fact.) Such densities are called multivariate totally positive of order 2 (= MTP_2). Suppose $\{p_\theta\}$ is a minimal standard exponential family whose dominating measure, ν, is MTP_2. It follows by the proof of the theorem above that then h non-decreasing implies $E_\theta(h)$ non-decreasing in θ.

Under suitable conditions it is also possible to derive analogous "order preserving" results for other partial orderings. For example, one may consider the partial ordering induced by a convex cone $C \subset R^k$, under which $x \propto_C y$ if $y - x \in C$.

A rather different but very fruitful partial ordering is that leading the notion of Schur convexity. Define $x \propto_S y$ if $\sum_{i=1}^{k} x_i = \sum_{i=1}^{k} y_i$ and if $\sum_{i=1}^{k'} x_{[i]} \leq \sum_{i=1}^{k'} y_{[i]}$, $1 \leq k' < k$, where $x_{[i]}$, $i=1,\ldots,k$, denote the coordinates of x written in decreasing order, etc. Then h is called Schur convex if it is non-decreasing relative to the ordering $\propto_S$. (Obviously any such function must be a symmetric function of $x_1,\ldots,x_k$.)

For further information about these and other partial orderings, consult Marshall and Olkin (1979), Karlin and Rinott (1981), Eaton (1982), and references cited in these works.

EXERCISES

2.2.1 Generalize 2.1(2) to

(1) $$||x||^{\ell} \frac{e^{\theta\cdot x} - e^{\theta_0\cdot x}}{||\theta - \theta_0||} \le K_{\ell+1} \sum_{i=1}^{I} e^{b_i\cdot x} .$$

Thus

(2) $$\frac{(\Pi x_i^{\ell_i})(e^{\theta\cdot x} - e^{\theta_0\cdot x} - (\theta - \theta_0) \quad x e^{\theta_0\cdot x})}{||\theta - \theta_0||} \le 2K_{\ell+1} \Sigma e^{b_i\cdot x} .$$

Use this to prove 2.2(1) by induction on ℓ.

2.3.1 Consider a one-dimensional standard exponential family with $K \subset [0, \infty)$. Show that

(1) $$(E_0[(1 - a)^X])^2 \ge E_0[(1 - 2a)^X], \quad 0 \le a < ½ ,$$

and $Var_0 X < \infty$ imply

(2) $$E_0(X) \ge Var_0 X .$$

[Let $e^{\theta} = (1 - a)$ and show by differentiating at $\theta = 0^-$ that (1) implies $\psi'(0^-) \ge \psi''(0^-)$. The finiteness of $Var_0 X$ guarantees that $\psi''(0^-) = Var_0 X < \infty$, etc., S. Zamir (personal communication).] (It is not known if (1) implies (2) without the assumption that $Var_0 X < \infty$.)

2.4.1 Canonical one-parameter exponential families for which $Var_{\theta}(X)$ is a quadratic function of $E_{\theta}(X)$ are called *quadratic variance function* families (= QVF). See Morris (1982, 1983). Verify that the following six families have the QVF property:

(1) $N(\mu, \sigma^2)$ μ known

(2) $P(\lambda)$

(3) $\Gamma(\alpha, \sigma)$ α known

(4) Bin (r, p) r known

(5) Neg. Bin. (r, p) r known

(6) ν has density $f(x) = (2 \cosh(\frac{\pi x}{2}))^{-1}$, $-\infty < x < \infty$,

relative to Lebesgue measure. ($X = \pi^{-1} \log(Y/(1 - Y))$ where $Y \sim$ Beta (½, ½).)

[In (6) $\psi(\theta) = -\log(\cos\theta)$. This is called the hyperbolic secant distribution. The generalized hyperbolic secant distributions are produced from these by infinite divisibility and convolution. These families are the only QVF families (Morris, 1982). See also Bar-Lev and Enis (1985).]

2.5.1 Let $\{p_\theta\}$ be a canonical one-dimensional exponential family. Then $N° = (\theta_1, \theta_2)$, $\xi(N°) = (\xi_1, \xi_2)$ for some $-\infty \le \theta_1 < \theta_2 \le \infty$ and $-\infty \le \xi_1 < \xi_2 \le \infty$. If $K = [x_1, \infty)$ then $\xi_1 = x_1$. (Theorem 3.6 is a multivariate generalization of this result.)

2.10.1 Let $\{p_\theta\}$ be a two-dimensional canonical exponential family. Find a convex subset of N such that h bounded and $E_\theta(h) = 0$ for all $\theta \in \partial N$ implies $h = 0$ a.e.(ν). (Hence, the family $\{p_\theta : \theta \in \partial N\}$ is "boundedly complete".) Conclude that every test of Θ_0 versus $\Theta_1 = N - \Theta_0$ is "admissible". (i.e. Let $\pi_\phi(\theta) = E_\theta(\phi)$. Then $\pi_{\phi_1}(\theta) \leqq \pi_{\phi_2}(\theta)$, $\theta \in \Theta_0$, and $\pi_{\phi_1}(\theta) \geqq \pi_{\phi_2}(\theta)$, $\theta \in \Theta_1$, implies $\pi_{\phi_1}(\theta) \equiv \pi_{\phi_2}(\theta)$.) [∂N contains an infinite number of line segments. See Farrell (1968).]

Similar Tests and Unbiased Tests

2.12.0 Let $\Theta_i \subset \Theta$, i=0,1. A critical test function ϕ, $0 \le \phi \le 1$, is called level α *unbiased* if $E_\theta(\phi) \le \alpha$, $\theta \in \Theta_0$, and $E_\theta(\phi) \ge \alpha$, $\theta \in \Theta_1$. It is called *similar* (level α) if $E_\theta(\phi) \equiv \alpha$, $\theta \in \bar{\Theta}_0 \cap \bar{\Theta}_1 \cap N$. The following problems consider the common case where $\Theta_0 \cup \Theta_1 = N$ so that $\partial\Theta_0 \cap N = \bar{\Theta}_0 \cap \bar{\Theta}_1 \cap N$. Exercises 2.21.3, 2.21.4 and 2.21.5 contain further applications of these concepts. See also 7.12.1.

2.12.1 Let $\{p_\theta\}$ be a regular canonical family and let $\theta' = (\theta'_{(1)}, \theta'_{(2)})$, $X' = (X'_{(1)}, X'_{(2)})$ be partitioned vectors. (Regularity is convenient but not essential here.) Let $L = \{\theta : \theta_{(1)} = 0\}$. Assume $L \cap N^o \ne \phi$.

(i) Show that a critical function ϕ is similar on L if and only if

$$\alpha = \int \phi(x)\nu(dx_{(1)}|x_{(2)}) \quad \text{a.e.}(\nu) \quad . \tag{1}$$

(Tests with property (1) are said to have *Neyman* structure. Note that the right side of (1) is $E_\theta(\phi|x_{(2)})$ for $\theta \in L$.)

(ii) Show that ϕ is similar on L and satisfies

$$(2) \qquad v \cdot \nabla E_\theta(\phi) = 0 \quad \forall \; v \in L^\perp = \{v : v \cdot \theta = 0 \quad \forall \; \theta \in L\}$$

if and only if ϕ satisfies (1) and

$$(3) \qquad \int x_{(1)}\phi(x)\nu(dx_{(1)}|x_{(2)}) = 0 \quad \text{a.e. } (\nu) .$$

(Note that (2) is a necessary condition for a test of H_0: $\theta \in L$ versus H_1: $\theta \notin L$ to be unbiased. (3) expresses the fact that $v \cdot \nabla E_\theta(\phi|x_{(2)}) = 0$ for all $\theta \in L$, $v \in L^\perp$, $x_{(2)}$. See Lehmann (1959) for many applications of (1) and (3) to the construction of U.M.P.U. tests.)

<u>2.12.2</u> (i) Let $X \sim N(\theta, I)$ in R^k, $k \geq 2$. Show there does not exist a non-constant level α similar test of $\Theta_0 = \{\theta : \theta_i \leq 0$ for some i$\}$. [Use Example 2.10.]

(ii) Show there exists a non-constant similar test of $\Theta_0 = \{\theta : \theta_i = 0$ for some i$\}$, but there does not exist a non-constant unbiased test of this hypothesis.

<u>2.12.3</u> Let $X \in R^k$, $X_i \sim P(\lambda_i)$, independent. Show there exists a non-trivial similar test of $\{\lambda : \lambda_i \leq 1 \; \forall \; i\}$ but there does not exist a non-trivial unbiased test of this hypothesis.

<u>2.13.1</u>

Let $X = (X_{ij})$ be a matrix $\Gamma(\alpha, I)$ variable. (See Exercise 1.14.4.) Observe that log $|X|$ has the same Laplace transform as $\sum_{i=1}^{m} \log Y_i$ where Y_i are independent $\Gamma(\alpha - (i-1)/2, 1)$ variables. Hence $|X|$ has the same distribution as $\prod_{i=1}^{m} Y_i$. Reinterpret this result to show equality of the distribution of the determinant of a Wishart (n, I) matrix and a product of independent χ^2 -variables.

<u>2.13.2</u>

Let F, G be two distributions on R^k. Let $\mu^F_{i_1,\ldots,i_k} = E(\prod_{j=1}^{k} X_j^{i_j})$ and similarly for μ^G. Suppose

(1) $$\mu^F_{i_1,\dots,i_k} = \mu^G_{i_1,\dots,i_k} \qquad i_j = 0,1,\dots \qquad j = 1,\dots,k ,$$

and

(2) $$\limsup_{n\to\infty} \frac{1}{n} m_{j,2n}^{\frac{1}{2}n} < \infty , \qquad j = 1,\dots,k$$

where $m_{j,2n} = E(|X_j|^{2n})$. (Note $m_{j,2n} = \mu_{0,\dots,2n,\dots,0}$.) Then $F = G$. (Condition (2) is slightly weaker than the necessary and sufficient condition,

(3) $$\sum_{n=1}^{\infty} m_{j,2n}^{-(\frac{1}{2}n)} = \infty , \qquad j = 1,\dots,k ,$$

for (1) to imply equality of F and G. See Feller (1966, Sections XV4 and VII3) and references cited therein.)

[Use Stirling's formula to show that $\Sigma m_{j,n}\theta^n/n!$ converges absolutely for $|\theta| < \varepsilon$, $j=1,\dots,k$, and hence that $\lambda_F = \lambda_G$ on an open set in R^k.]

2.14.1 (Bar-Lev (1983).)

Let $X \sim \text{Expf}\ (\theta)$ with $\Theta^\circ \neq \phi$. Let $\Sigma(X_{(2)}|x_{(1)})$ denote the indicated conditional covariance matrix. Show that $\Sigma_\theta(X_{(2)}|x_{(1)})$ depends only on θ if and only if $X_{(1)} \perp (X_{(2)} - h(X_{(1)}))$ for some (measurable) function h.

[Integrate $\Sigma_\theta(X_{(2)}|x_{(1)})$ on θ starting at $0 \in \Theta^\circ$ to find that the conditional cumulant generating function of $X_{(2)}$ under P_0 is

(2) $$\psi(\theta|x_{(1)}) = \rho(\theta_{(2)}) + \theta_{(2)} \cdot h(x_{(1)}) \quad \text{for some functions } \rho , h .$$

Show that (2) implies $X_{(2)} - h(X_{(1)}) \perp X_{(1)}$ under P_0.]

2.14.2

Suppose $X \sim \text{Expf}\ (\theta)$ with $\Theta^\circ \neq \phi$. Then the following are equivalent:

(1) $X_{(1)} \perp X_{(2)}$ for some $\theta^\circ \in \Theta$, or for all $\theta \in \Theta$,

(2) $\psi(\theta_{(1)}, \theta_{(2)}) = \psi_1(\theta_{(1)}) + \psi_2(\theta_{(2)})$ for some functions ψ_1 and ψ_2 ,

(3) $X_{(i)} \sim \text{Expf}\ (\theta_{(i)})$ for i = 1 and 2 ,

(4) $\text{cov}_\theta(X_{(1)}, X_{(2)}) = 0 \quad \forall\, \theta \in \Theta$.

[For (1) - (3) apply Theorem 2.14 with $h \equiv 0$ and check $\psi_i = \rho_i$, $i=1,2$. For (4) $\Rightarrow$ (2) use 2.3(2) and integrate.]

2.14.3 (Patil (1965), Barndorff-Nielsen and Blaesild (1983).)

Let $\mathcal{P} = \{P_\theta : \theta \in \Theta\}$ be a family of distributions on $\mathcal{Y}$, $\mathcal{B}$. Let $X : \mathcal{Y} \to R^k$ (measurable), $\Theta \subset R^k$ with $\Theta^\circ \neq \phi$. Suppose

$$\ln E_\theta \exp((\beta - \theta) \cdot X(Y)) = \rho(\beta) - \rho(\theta), \qquad \beta,\theta \in \Theta$$

for some function $\rho(\cdot)$. Then $X \sim \text{Expf}\,(\theta)$. [Use Corollary 2.13.]

2.14.4

Let X have a k-dimensional multinomial (N, π) distribution. Write $X'_{(1)} = (X_1,\ldots,X_{k_1})'$, $X'_{(2)} = (X_{k_1+1},\ldots,X_k)'$. Show that the marginal distributions of both $X_{(1)}$ and $X_{(2)}$ form an exponential family, but $X_{(1)}$ is not independent of $X_{(2)}$ as one might expect from Theorem 2.14(2). Why not? [The fact that X is not a minimal family is irrelevant; for $k \geq 3$, $k_1 \leq k-2$ the same phenomenon occurs in the minimal model defined as in 1.2(7).]

2.15.1

Let the independent symmetric m×m matrices, X_i, $i=1,\ldots,n$, have matrix $\Gamma(\alpha_i, \not{\Sigma})$ distributions. (See Exercise 1.14.4). Show that $Z = Z_1,\ldots,Z_n$ with $Z_j = |X_j|/|\sum_{i=1}^n x_i|$ is independent of $\sum_{i=1}^n x_i$. Show that the distributions of $\ln Z = \{\ln Z_j : j=1,\ldots,n\}$ form an exponential family, and identify the canonical statistic and parameter for this distribution. (This generalizes Example 2.15(ii). The distributions of Z form the so-called multivariate beta distribution. See, e.g., Muirhead (1982). When $m = 1$ the X_i have ordinary Γ distributions and the distribution of Z is a Dirichlet distribution. See Exercise 5.6.2.

2.16.1

Suppose $\nu_n \to \nu$ with $\nu(R^k) < \infty$. Then

(1) $$\limsup \nu_n(R^k) \leq \nu(R^k)$$

if and only if the sequence $\{\nu_n\}$ is tight. [Let $c(x) = i$ if $r_i \leq ||x|| \leq r_{i+1}$

and choose $r_i \ni \sup_n \nu_n(\{||x|| \geq r_i\}) \leq 1/i^2$, i=1,2,... .] Hence a convergent sequence of probability measures has a probability measure as its limit if and only if it is tight.

2.17.1

Verify 2.17(4),(5). [From Lemma 2.1(1)

(1) $$||\int xe^{b\cdot x}\ \nu_i(dx)|| \leq \sum_{j=1}^{J} ||\int \left\{ \frac{xe^{b\cdot x}}{\Sigma e^{b_j\cdot x}} \right\} e^{b_j\cdot x}\ \nu_i(dx)||$$

and the quantity in braces in (1) is $O(1/(1 + ||x||))$. Now use 2.16(1).]

2.17.2

Let $S \subset R^k$ and $B = \text{conhull}\ S$. Let ν_n be a bounded sequence of measures on R^k ($\nu_n(R^k) \leq K_1 < \infty$) with $\lambda_{\nu_n}(b) < \infty$, $b \in S$, n=1,... . Suppose $0 \in B^\circ$. Define $P_{n,b}$ by

(1) $$\frac{dP_{n,b}}{d\nu_n} = \exp(b\cdot x - \psi_{\nu_n}(b)) \quad .$$

Suppose for each $b \in S$ there is a $K = K(b)$ such that

(2) $$\limsup_{n\to\infty} P_{n,b}(\{||x|| < K\}) > 0 \ .$$

Then there is a subsequence $\{n'\} \subset \{n\}$ and a non-zero limiting measure ν such that for all $b \in B^\circ$

(3) $$e^{b\cdot x}\ \nu_{n'}(dx) \to e^{b\cdot x}\ \nu(dx)\ , \qquad \lambda_{\nu_{n'}}(b) \to \lambda_\nu(b)$$

[As in the proof of Theorem 2.17 it suffices to consider the case where S is finite. Then $K = \max\{K(b) : b \in S\} < \infty$. If $b \in S$, $||b|| \leq K_0$ then $0 < \varepsilon \leq \int_{||x||\leq K} e^{b\cdot x}\nu_{n'}(dx)/\lambda_{\nu_{n'}}(b) \leq K_1 e^{K_0 K}/\lambda_{\nu_{n'}}(b)$. Hence 2.17(1) is satisfied on $S \cap \{b : ||b|| \leq K_0\}$. $\nu \neq 0$ since $0 \in B^\circ$]

2.18.1

Let $\{\nu_{\alpha n} : \alpha \in A\}$, n=1,2,... be a family of sequences of measures on $X = \{0,1,\ldots\}$. Show that $\nu_{\alpha n} \to \nu_\alpha$ uniformly in α if and only if

$\nu_{\alpha n}(\{x\}) \to \nu_\alpha(\{x\})$ uniformly in α for each $x \in X$.

2.19.1

Let $\{p_\theta\}$ be an exponential family with supp $\nu \subset \{0,1,\ldots\}$ and $\nu(0) > 0$, $\nu(1) > 0$. Let $X_1,\ldots,X_n$ be a random sample and, as usual, let $S_n = \sum_{i=1}^{n} X_i$. Define $\theta_n(\lambda)$ by

$$(1) \qquad \xi(\theta_n(\lambda)) = \lambda/n \quad .$$

Let $F_{\lambda,n}$ denote the distribution of S_n under the parameter $\theta_n(\lambda)$. Show that $F_{\lambda,n} \to P(\lambda)$ and the convergence is uniform in λ over $\lambda \in [a,b]$ for $0 < a < b < \infty$. (A slight elaboration of the argument yields uniformity over $[0, b]$.) Generalize this result to the case where p_θ is a k-dimensional exponential family. [Show $\psi''(\theta_n(\lambda)) \to 0$ as $n \to \infty$ since $\theta_n(\lambda) \to -\infty$, uniformly for $\lambda \in [a, b]$. Hence $\log E_{\theta_n(\lambda)} e^{\beta S_n} = \lambda(e^\beta - 1) + o(1)$ as $n \to \infty$ uniformly for $\lambda \in [a, b]$. Then apply Theorem 2.19. In the non-degenerate k-dimensional case the limit distribution is the product of independent Poisson variables.] (A special case of the above is the well known result Bin $(n, \lambda/n) \to P(\lambda)$. The general form of the above statement was pointed out to me by I. Johnstone.)

2.21.1

Let X be non-central χ^2 with m degrees of freedom and non-centrality parameter θ . Show that the distributions of X have the sign-change preserving properties 2.21(2), (3). [Use Exercise 1.12.1(1). Write $E_\theta(h(X)) = E_\theta(E(h(X)|K))$.]

2.21.2

Let X be a one-dimensional exponential family and $\theta_0 \in N°$.

(i) Show that the (essentially unique) level α test of the form

$$(1) \qquad \phi(x) = \begin{cases} 1 & x > x_0 \\ \gamma & x = x_0 \\ 0 & x < x_0 \end{cases}$$

is the U.M.P. level α test of H_0: $\theta \leq \theta_0$ versus H_1: $\theta > \theta_0$.

(ii) Similarly, show that the (essentially unique) level α test of the form

$$\phi(x) = \begin{cases} 1 & x > x_2 \text{ or } x < x_1 \\ \gamma_i & x = x_i \\ 0 & x_1 < x < x_2 \end{cases} \tag{2}$$

satisfying

$$E_{\theta_0}(x\phi(x)) = 0 \tag{3}$$

is the U.M.P.U. level test of H_0: $\theta = \theta_0$ versus H_1: $\theta \neq \theta_0$.

[(i) Let ϕ' be any different level α test. Then $S^-(\phi - \phi') = 1$. $E_{\theta_0}(\phi - \phi') = 0$ by definition. Now use Theorem 2.18. (ii) Condition (3) is the one-dimensional version of 2.12.1(3). Again use Theorem 2.18.] (It is also possible to show by a continuity argument that level α tests of the form (1) and (2), (3) always exist.)

2.21.3

Consider a 2×2 contingency table. (See Exercise 1.8.1.) Describe the general form of the U.M.P.U. level α tests of the following null hypotheses. In each case the alternative is the complement of H_0.

(i) H_0: $p_{11}p_{22}/p_{12}p_{21} \leq 1$

(ii) H_0: $p_{11}p_{22}/p_{12}p_{21} = 1$

(iii) H_0: $p_{11} \leq p_{12}$

(iv) H_0: $p_{12} = p_{21}$.

(This corresponds to the exact form of McNemar's test. See, e.g. Fleiss (1981).) [Use Exercise 2.21.2 and, for (i), (ii), Exercise 1.15.1. See Lehmann (1959).]

2.21.4

Consider a 2×2 contingency table. Let $c > 0$, $c \neq 1$. Show there exist non-trivial similar tests of the null hypothesis

H_0: $p_{11}/(p_{11} + p_{12}) = cp_{21}/(p_{21} + p_{22})$ of conditional probabilities in a given proportion, even though this is not a log-linear hypothesis. [Use randomized tests. Consider the conditional distribution given Y_{i+}, $i=1,2$ under which Y_{11} and Y_{21} are independent binomials. (This case is of interest on its own merits.) Consider the special case $Y_{1+} = 1 = Y_{2+}$ for which the condition for similarity reduces to four linear equations in the four variables $\phi(y)$ for the four conditionally possible outcomes, y. This test is unbiased for the one-sided version of H_0, but not for H_0 as defined above. Is there, in general, an unbiased test of H_0? Is there, in general, a U.M.P.U. test of either the one- or two-sided hypothesis in either the original model or the conditional (independent binomial) model? The somewhat analogous question of the existence of similar and of unbiased tests for the Behrens-Fisher problem of equality of means for two normal samples with unknown variances is solved in Wijsman (1958) and in Linnik (1968).]

2.21.5

Let $X_1,\ldots,X_n$ be a sequence of independent failure times, assumed to have a $\Gamma(\alpha, \sigma)$ distribution. Describe the U.M.P.U. tests of H_0: $\alpha = 1$ versus H_1: $\alpha > 1$ and H_1': $\alpha \neq 1$. [Use Exercise 2.21.2 and Example 2.15.]

2.25.1

Suppose ν has density f with respect to Lebesgue measure on R^k and f is MTP_2 (i.e. has monotone likelihood ratio) in each pair of coordinates. Prove the conclusions of Lemma 2.24 and Theorem 2.25. Prove these also for the case where f, as above, is a density with respect to counting measure on the lattice of points with integer coordinates. [If $h(x_1,\ldots,x_k)$ is non-decreasing then, under ν, $E(h(X_1,\ldots,X_{k-1}, X_k) | X_k = x_k)$ is also non-decreasing.]

2.25.2

Let $\{p_\theta\}$ be a canonical k-parameter exponential family with

$\theta_0 \in N^\circ$. Let H_0: $\theta \le \theta_0$ and H_1: $\theta > \theta_0$. (i) Show that any Bayes or generalized Bayes test, α, of H_0 versus H_1 has the strong monotonocity property

$$(1) \qquad \begin{array}{lll} \phi(x) > 0 & y > x & \Rightarrow \phi(y) = 1 \\ \phi(x) < 1 & y < x & \Rightarrow \phi(y) = 0 \end{array} .$$

Assume $\theta_0 = 0$ and consider $\nabla \int p_\theta(x)[G_1(d\theta) - G_0(d\theta)]$ where G_i denotes the (generalized) prior measure restricted to H_i.] (ii) Suppose the dominating measure ν is MTP_2. Show that any (generalized) Bayes test is unbiased. [Use the above and Exercise 2.25.1.]

2.25.3 (Slepian's Inequality)

Let X, Y be k-dimensional normal variables with mean 0 and non-singular covariance matrices A, B, respectively. Suppose

$$a_{ii} = b_{ii}, \quad a_{ij} \ge b_{ij} \qquad 1 \le i,j \le k \quad .$$

Then, for any $C \in R^k$,

$$(1) \qquad \Pr\{X \le C\} \ge \Pr\{Y \le C\} \quad .$$

[If $Z^{(\rho)} \sim N(0, A + \rho(B - A))$ then

$$(2) \qquad \frac{\partial}{\partial\rho} P(Z^{(\rho)} \le C) = \sum_{i \ne j} \alpha_{ij} \frac{\partial}{\partial\theta_{ij}} P(Z^{(\rho)} \le C)$$

where each $\alpha_{ij} \ge 0$. Note that for $i \ne j$

$$(3) \qquad \frac{\partial\lambda}{\partial\theta_{ij}} = \theta_{ij} \exp(-\ln|\Sigma|/2) = \theta_{ij} \lambda$$

by 2.4(2). Hence

$$(4) \qquad \frac{\partial p_\theta(Z)}{\partial\theta_{ij}} = \theta_{ij}\, p_\theta(Z) = \frac{\partial^2 p_\theta(Z)}{\partial Z_i\, \partial Z_j}$$

from Corollary 2.13. Combine (2) and (4) to yield (1).] (For an alternate proof of Slepian's inequality see Saw (1977). For generalizations see Joag-Dev, Perlman, and Pitt (1983) and Brown and Rinott (1986).)

CHAPTER 3. PARAMETRIZATIONS

In regular exponential families maximum likelihood estimation is closely related to the so-called mean value parametrization. This parametrization will be described after some brief preliminaries. The relation to maximum likelihood is pursued in Chapter 5.

3.1 Notation

For $v \in R^k$, $\alpha \in R$ let $H(v, \alpha)$ denote the hyperplane

$$H(v, \alpha) = \{x \in R^k : v \cdot x = \alpha\} .$$

Let $H^+(a, \alpha)$ and $H^-(a, \alpha)$ be the open half spaces

$$H^+(v, \alpha) = \{x \in R^k : v \cdot x > \alpha\}$$

$$H^-(v, \alpha) = \{x \in R^k : v \cdot x < \alpha\} .$$

When (v, α) are clear from the context they will be omitted from the notation. Note that the closure of $H^\pm$ is written $\overline{H}^\pm$ and, of course, satisfies $\overline{H}^\pm = H \cup H^\pm$.

STEEP FAMILIES

Most exponential families occurring in practice are regular (i.e. N is open). However, for technical reasons which will become clear in Chapter 6, it is very useful to prove the parametrization Theorem 3.6 for steep families as well.

3.2 Definition

Let $\phi: R^k \to (-\infty, \infty]$ be convex. Let $N = \{\theta \in R^k: \phi(\theta) < \infty\}$. Assume ϕ is continuously differentiable on N°. Let $\theta_1 \in N - N^\circ$, $\theta_0 \in N^\circ$, and let $\theta_\rho = \theta_0 + \rho(\theta_1 - \theta_0)$, $0 < \rho < 1$, denote points on the line joining θ_0 to θ_1. Then, ϕ is called *steep* if for all $\theta_1 \in N - N^\circ$, $\theta_0 \in N^\circ$,

(1) $$\lim_{\rho\uparrow 1} (\theta_1 - \theta_0) \cdot \nabla\phi(\theta_\rho) = \infty \quad .$$

Note that (1) is the same as

(1') $$\lim_{\rho\uparrow 1} \frac{\partial}{\partial\rho} \phi(\theta_\rho) = \infty \quad .$$

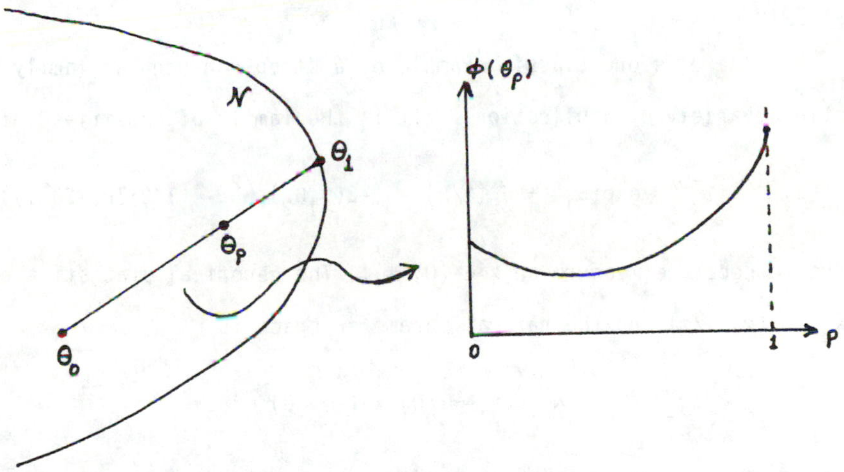

Figure 3.2(1): An illustration of the definition of steepness

A standard exponential family is called steep if its cumulant generating function, ψ, is steep. (A steep convex function is sometimes referred to as an "essentially smooth" convex function.) Note that if the exponential family is regular then it is *a fortiori* steep since $N - N^\circ = \phi$.

Here is a convenient necessary and sufficient condition for steepness.

3.3 Proposition

A minimal standard exponential family is steep if and only if

(1) $$E_\theta(||x||) = \infty \quad \text{for all} \quad \theta \in N - N^\circ \quad .$$

Proof. Suppose the family is steep. Then

$$(\theta_1 - \theta_0) \cdot \nabla\psi(\theta_\rho) = (\theta_1 - \theta_0) \cdot \xi(\theta_\rho) \to \infty \quad \text{as} \quad \rho \uparrow 1 \quad .$$

This implies $E_{\theta_\rho}((\theta_1 - \theta_0) \cdot X) \to \infty$, which implies (1).

The converse seems not to be easy to prove without further preparation. We postpone the proof to Chapter 6. It appears after the proof of Lemma 6.8. ||

3.4 Example

There is one classic example of a steep non-regular family which occurs in a variety of applications. It is the family of densities defined by

(1) $$(\pi)^{-1/2} z^{-3/2} \exp(\theta_1 z + \theta_2(1/z) - (-2(\theta_1\theta_2)^{1/2} - (1/2)\ln(-2\theta_2)))$$

relative to Lebesque measure on $z \in (0, \infty)$. The canonical statistics are $(x_1, x_2) = (z, 1/z)$ and the natural parameter space is

(2) $$N = (-\infty, 0] \times (-\infty, 0) \quad .$$

Thus the family is not regular but is steep since $E_{(0,\theta_2)}(x_1) = \infty$ for all $\theta_2 \in (-\infty, 0)$. These densities are referred to as *inverse Gaussian.* They arise, for example, as the distribution of the first time (x_1) that a standard Brownian motion crosses the line $\ell(t) = \sqrt{-2\theta_2} - \sqrt{-2\theta_1}\, t$. Note that these densities with $\theta_1 = 0$ are the scale family of stable densities on $(0, \infty)$ with index ½. See Feller (1966). For some other steep non-regular families see Bar-Lev and Enis (1984).

MEAN VALUE PARAMETRIZATION

We begin with a useful lemma which involves a natural relation between parameter space (Θ) and sample space (X). Similar relations will reoccur several times and we have found it useful to draw pictures to illustrate the geometric relationships involved. Figure 3.5.1, below, is a simple example of such a picture which illustrates the hypotheses of Lemma 3.5.

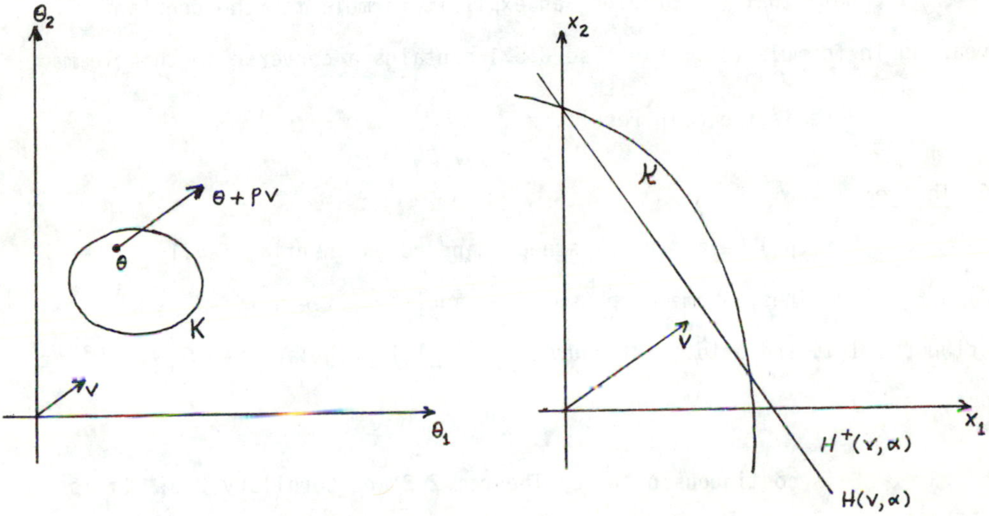

Figure 3.5.1: Illustrating the hypotheses of Lemma 3.5 when $k = 2$.

3.5 Lemma

Let $v \in R^k$, $\alpha \in R$. Let $K \subset R^k$ be compact. Suppose $\nu(\bar{H}^+(v, \alpha)) > 0$. Then there exists a constant $c > 0$ such that

(1) $$\lambda(\theta + \rho v) \geq ce^{\rho\alpha} \qquad \forall\, \theta \in K, \qquad \rho \geq 0 \quad .$$

(Note that (1) is equivalent to

(1') $$\psi(\theta + \rho v) \geq \rho\alpha + \log c \qquad \forall\, \theta \in K, \qquad \rho \geq 0$$

If $\theta + \rho v \notin N$ then $\lambda(\theta + \rho v) = \infty$ so that (1) is trivial.)

Proof.

(2) $$\lambda(\theta + \rho v) = \int e^{(\theta+\rho v)\cdot x}\nu(dx) \geq e^{\rho\alpha} \int_{\bar{H}^+} e^{\theta\cdot x}\nu(dx) \geq ce^{\rho\alpha}$$

where

(3) $$c = \inf_{\theta\in K} \int_{\bar{H}^+} e^{\theta\cdot x}\nu(dx) > 0 \quad .$$

((2) shows that if $c = \infty$ here then $\lambda(\theta + \rho v) = \infty$ for all $\theta \in K$ and all $\rho \geq 0$.) ||

Note that (3) provides an explicit formula for the constant c appearing in formula (1). Exercise 3.5.1 contains a converse to this lemma.

Here is the main result.

3.6 Theorem

Let $\{p_\theta\}$ be a minimal steep standard exponential family. Then $\xi(\theta) = E_\theta(X)$ defines a homeomorphism of N° and K° (i.e., $\xi: N^\circ \to K^\circ$ is continuous, 1-1, and onto. Of course, if $\{p_\theta\}$ is regular then $\xi: N \to K^\circ$ since $N = N^\circ$).

Proof. ξ is continuous on N° by Theorem 2.2 and Corollary 2.3. It is 1-1 by Corollary 2.5. It remains to prove that $\xi(N^\circ) = K^\circ$, that is, to show

(1) $$x \in K^\circ \Rightarrow x \in \xi(N) \quad .$$

It suffices to prove (1) for $x = 0$, for then the desired result for arbitrary $x \in K^\circ$ follows upon translating the origin, which is justified by Proposition 1.6. So, assume $0 \in K^\circ$.

Let $S_1 = \{v \in R^k: ||v|| = 1\}$. Since $0 \in K^\circ$ there is an $\varepsilon > 0$ such that

(2) $$\nu(\bar{H}^+(v, \varepsilon)) > c > 0$$

for all $v \in S_1$. (If not, there would be sequences $v_i \in S_1$ with

$v_i \to v \in S_1$ and $\varepsilon_i \to 0$ for which $\nu(\bar{H}^+(v_i, \varepsilon_i)) \to 0$. This would imply $\nu(\bar{H}^+(v, 0)) = 0$ which contradicts $0 \in K°$.) Now apply Lemma 3.2 (with $v = \theta/||\theta||$ and $\rho = ||\theta||$) including the expression 3.2(3) for the constant appearing in the lemma to get

(3) $$\psi(\theta) \geq ||\theta||\varepsilon + \log c$$

with c as in (2). Thus

(4) $$\lim_{||\theta|| \to \infty} \psi(\theta) = \infty \quad .$$

(See Exercise 3.6.2 and Lemma 5.3(3) for restatements of (3), (4).)

Any lower semi-continuous function (such as ψ) defined on a closed set and which also satisfies (4) must assume its minimum. To see this, let $\psi(\theta_i) = \inf \{\psi(\theta) : \theta \in R^k\}$. $||\theta_i|| \to \infty$ is impossible by (4). So, there is a convergent subsequence, $\theta_{i'} \to \theta^*$, and $\psi(\theta^*) = \inf \{\psi(\theta) : \theta \in R^k\}$ by lower semi-continuity.) This minimum is assumed at a point $\theta^* \in N$.

Suppose $\theta^* \in N - N°$. Then, for some $0 < \rho' < 1$, $\psi(\theta_{\rho'}) < \psi(\theta^*) = \lim_{\rho \uparrow 1} \psi(\theta_0 + \rho(\theta^* - \theta_0))$ by virtue of 3.2(1') of the definition of steepness. Hence no $\theta^* \in N - N°$ can be the minimum point for ψ. It follows that $\theta^* \in N°$.

Hence

$$\xi(\theta^*) = \nabla\psi(\theta^*) = 0$$

since ψ is differentiable on a neighborhood of θ^*. (Here we use Theorem 2.2, Corollary 2.3, and the fact that $\theta^* \in N°$ an open set.) This proves (1) for $x = 0$ and, as noted, completes the proof of the theorem. ||

3.7 Interpretation

Theorem 3.6 shows that a minimal, steep family with parameter space $N°$ can be parametrized by $\xi = \xi(\theta)$, and the range of this parameter is $K°$. This is the *mean value parametrization*. In this parametrization the resulting family is an exponential family, but of course is no longer a standard exponential family (except when $\xi(\cdot)$ is affine). Write

$$(1) \qquad \theta(x) = \xi^{-1}(x) = (\theta : \xi(\theta) = x) \quad .$$

The exponential family parametrized by ξ then has densities $\hat{p}_\xi(x) = \exp(\theta(\xi) \cdot x - \psi(\theta(\xi)))$. For a number of applications this parametrization is more convenient than the "natural" parametrization described by the canonical parameter θ. If $\{p_\theta\}$ is regular then $N = N^\circ$ and the mean value parametrization reparametrizes the full family.

Minimality was used in Theorem 3.6 only to guarantee that the map is 1-1. Even without minimality the map ξ discriminates between different distributions in $\{P_\theta: \theta \in N\}$. Hence one can still use the mean-value parametrization to conveniently index $\{P_\theta: \theta \in N^\circ\}$, and the range of the mean value parameter is the relative interior of K. (Equivalently, one may reduce to a minimal family by Theorem 1.9 and then apply Theorem 3.3.)

If the family is not steep then $\xi(N^\circ) \subset K^\circ$. We leave this fact -- relatively unimportant for statistical application -- as an exercise. In this case it is even possible to have $\xi(N^\circ)$ not convex. See Exercise 3.7.1 for an example due to Efron (1978).

3.8 Example (Fisher-VonMises Distribution)

For a number of common exponential families the mean value parametrization is the familiar parametrization, or nearly so. For example, for the Binomial (N, π) family the expectation parameter is $N\pi$, for the Poisson (λ) family the expectation parameter is λ, and for the exponential distributions (gamma distributions with index $\alpha = 1$ and unknown scale, σ) the expectation parameter is σ. For the multivariate normal (μ, Σ) family the expectation parameters are μ and $\mu\mu' + \Sigma$ (corresponding to the canonical statistics of 1.14). The mean value parameters are not always so convenient. Nevertheless it is necessary to consider this parametrization in order to construct maximum likelihood estimators. See especially Theorem 5.5.

Accordingly, we now discuss the mean value parametrization for the Fisher-VonMises distribution.

Let ν be uniform measure on the sphere of radius one in R^k. Consider the exponential family generated by ν. When $k = 2$ this is the *VonMises family*. When $k = 3$ it is the *Fisher family* of distributions. These distributions appear often in applications, with a variety of parametrizations, to model angular data in R^k. Consult Mardia (1972) for an extended treatment of these families; see also Beran (1979). (Frequently one considers a sample of n observations from one of these distributions. The sample mean, $\bar{X}_n$, is then also said to have a VonMises or Fisher distribution. The mean value parametrization for the family of distributions of $\bar{X}_n$ is, of course, identical to that below since $E_\theta(\bar{X}_n) = E_\theta(X)$. See also 5.5(3).)

The Laplace transform of ν is

(1) $$\lambda_\nu(\theta) = (2\pi)^{k/2}\, I_{k/2-1}(||\theta||)/||\theta||^{k/2-1}$$

where $I_s(\cdot)$ denotes the modified Bessel function of order s. When k is odd these functions have a convenient representation in terms of hyperbolic functions; for example

(2) $$I_{1/2}(r) = (2/\pi r)^{1/2} \sinh r$$

$$I_{3/2}(r) = (2/\pi r)^{1/2}(\cosh r - (\sinh r)/r) \quad .$$

(See, for example, Courant and Hilbert (1953).) These functions also have nice recurrence relations; in particular

(3) $$I_s'(r) = I_{s+1}(r) + sI_s(r)/r\ , \qquad s \geq 0, \qquad r > 0 \quad .$$

By symmetry, or by calculation, it follows that $\xi(\theta)$ lies in the same direction as θ, that is

(4) $$\xi(\theta)/||\xi(\theta)|| = \theta/||\theta||\ , \qquad \theta \neq 0, \quad \text{and} \quad \xi(0) = 0 \quad .$$

It remains therefore to give a formula for $||\xi(\theta)||$. For this purpose it suffices to consider the case where $\theta_r = (r,0,\ldots,0)$, and to calculate $\frac{d}{dr}\ell n\,\lambda_\nu(\theta_r)$. For the Fisher distribution (k = 3) one gets from (1) - (3) that

(5) $$||\xi(\theta)|| = \coth ||\theta|| - ||\theta||^{-1} .$$

For the Von Mises distribution (k = 2) one gets only the less convenient expression

(6) $$||\xi(\theta)|| = I_1(||\theta||)/I_0(||\theta||) .$$

Although (6) is less convenient that (5), it can be used in conjunction with series expansions or tables of the modified Bessel function to provide numerical values for $||\xi(\theta)||$, and other information about $||\xi(\theta)||$.

MIXED PARAMETRIZATION

We refer to the type of situation discussed in 1.7. $M = \binom{M_1}{M_2}$ is a partitioned k×k non-singular matrix with $M_1M_2' = 0$. Write

(1) $$M_i x = z_i \qquad i = 1, 2$$
$$(M_i^-)'\theta = \phi_i \qquad i = 1, 2 .$$

(Thus $\binom{\phi_1}{\phi_2} = (M^{-1})'\theta$.) Where convenient we write $\phi_i = \phi_i(\theta)$ to emphasize the dependence on θ, etc.)

Note that

(2) $$M_i\xi(\theta) = E_\theta(M_iX) = E_\theta(Z_i) = \zeta_i(\theta) \quad \text{(say)} \quad i = 1, 2 .$$

Recall also that one may without loss of generality visualize only the case where M = I. In this case $\phi_1' = (\theta_1,\ldots,\theta_m)$, $z_2' = (x_{m+1},\ldots,x_k)$, $\zeta_2' = (\xi_{m+1},\ldots,\xi_k)$, etc.

The following result is valid for steep families but for simplicity we state and prove it here only for regular families. See Exercise 3.9.1.

3.9 Theorem

Let $\{p_\theta\}$ be minimal and regular. Then the map

(3) $$\theta \to \begin{pmatrix} \zeta_1(\theta) \\ \phi_2(\theta) \end{pmatrix}$$

is 1 - 1 and continuous on N° $(=N)$ with range

(4) $$\zeta_1(N^\circ) \times \phi_2(N^\circ) = K^\circ_{(1)} \times \phi_2(N^\circ) \quad .$$

Proof. Fix $\phi_2^0 \in \phi_2(N)$ and refer to Theorem 1.7. The distributions of Z_1 given $\phi_2(\theta) = \phi_2^0$ form the minimal regular standard exponential family generated by $\nu_{\phi_2^0}$. According to Theorem 3.6 this family can be parametrized (in a 1 - 1 manner) by $\zeta_1(\theta) = E_\theta(Z_1)$. The range of this map is

$$\text{int (conhull (supp } \nu_{\phi_2^0})) = K^\circ_{\phi_2^0} \qquad \text{(say)} \quad .$$

The formula for $\nu_{\phi_2^0}$ is given in 1.7(5), but all that needs to be noted is that $K^\circ_{\phi_2^0} \equiv K^\circ_{(1)}$. The map in (3) is therefore 1 - 1 with range as in (4). Continuity of the map in (3) is immediate from continuity of ξ. ||

3.10 Interpretation

The above theorem has an interpretation like that of Theorem 3.6. Any minimal regular exponential family can be parametrized by parameters of the form 3.9(3), above. This parametrization is called the *mixed parametrization.*

Consider a mixed parametrization with parameter $\binom{\zeta_1}{\phi_2}$, as above. Then the family of densities corresponding to the parameters $\{\binom{\zeta_1}{\phi_2} : \phi_2 = \phi_2^0\}$

forms a full standard exponential family of order m. (See Theorem 1.7.) However, if one fixes the expectation coordinate and looks at the family corresponding to the parameters $\{\binom{\zeta_1}{\phi_2} : \zeta_1 = \zeta_1^0\}$ then one gets in general only some non-full standard family of dimension and order k, whose parameter space is a (k - m) dimensional manifold in N. Here is an example.

Consider the parametrization of the three dimensional multinomial (N, π) family discussed following 1.8(6). A mixed parametrization for this family involves

$$\binom{\zeta_1}{\zeta_2} = E\binom{Z_1}{Z_2} = \binom{2\pi_1 + \pi_2}{\pi_2 + 2\pi_3} N$$

and

$$\phi_3 = (\tfrac{1}{2}) \log (\pi_2^2/4\pi_1 \pi_3) \quad .$$

Note that the range of $\binom{\zeta_1}{\zeta_2}$ is

$$\{\binom{\zeta_1}{2N - \zeta_1} : 0 < \zeta_1 < 2N\}$$

independent of the value of $\phi_3 \in (-\infty, \infty)$, as claimed by Theorem 3.9. For fixed $\phi_3 = \phi_3^0$ the distributions of $\binom{Z_1}{Z_2}$ form a 2 dimensional exponential family (of order 1) having expectation parameter $\binom{\zeta_1}{\zeta_2}$. (In the genetic interpretation for this parametrization the parameter ϕ_3 measures the strength of selection in favor of the heterozygote character Gg.)

On the other hand the family of distributions corresponding to fixed $\binom{\zeta_1}{\zeta_2}$ is not so convenient. It is the non-linear subfamily of the usual full standard family described by

(1) $$\Theta = \{\theta : 2e^{\theta_1} + e^{\theta_2} = (\zeta_1/N)\Sigma e^{\theta_i}\} \quad .$$

(If one reduces the usual standard exponential family to a minimal family of dimension 2, then the parameter set becomes a smooth one-dimensional curve within R^2. This provides an example of a curved exponential family, as defined below. See Exercise 3.11.2.)

DIFFERENTIABLE SUBFAMILIES

3.11 Description

A *differentiable subfamily* is a standard exponential family with parameter space Θ an m-dimensional differentiable manifold in N. An especially convenient situation occurs when Θ is a one-dimensional manifold -- i.e. a differentiable curve. Such a family is called a *curved exponential family*. (A technical point: it is often convenient to assume that the parameter space is smoother than being merely differentiable -- for example, to assume it possesses second derivatives. Whenever convenient we consider such an assumption *implicit* in the definition of a differentiable subfamily; writing formulae for relevant second or higher derivatives (as in (3) below) carries with it the assumption that these derivatives exist.)

In a differentiable subfamily the parameter space can be written locally as $\{\theta(t) : t \in N\}$ where N is a neighborhood in R^m and $\theta(\cdot)$ is differentiable and one to one. Properties of such a family around some $\theta_0 \in \Theta$ can often be most conveniently studied after invoking Proposition 1.6 to rewrite the family in a more convenient form. For example in a curved exponential family m = 1 and the proper choice of ϕ_0, z_0 and M in that proposition transforms the problem into one in which

$$\theta_0 = 0 = \theta(t_0)$$

(1) $$\xi(\theta_0) = E_{\theta_0}(X) = 0$$

$$\not{\Sigma}(\theta_0) = I$$

(2)
$$\dot\theta(t_0) = \frac{d}{dt}\theta(t_0) = \begin{pmatrix} a \\ 0 \\ \vdots \\ 0 \end{pmatrix}$$

$$\ddot\theta(t_0) = \frac{d^2}{dt^2}\theta(t_0) = \begin{pmatrix} a^2 b \\ a^2/\rho \\ 0 \\ \vdots \\ 0 \end{pmatrix} .$$

(The value $\rho = \infty$ is possible.) Furthermore, one can linearly reparametrize the curve so that $\theta_0 = \theta(0)$ (i.e. so that $t_0 = 0$) and so that $a = 1$ and (2) becomes

(3)
$$\frac{d}{dt}\theta(0) = \begin{pmatrix} 1 \\ 0 \\ \vdots \\ 0 \end{pmatrix} , \qquad \frac{d^2}{dt^2}\theta(0) = \begin{pmatrix} b \\ 1/\rho \\ 0 \\ \vdots \\ 0 \end{pmatrix} .$$

In this form ρ is the radius of curvature of the curve $\theta(t)$ at $t = 0$. The value of $1/\rho$ is sometimes referred to as the *statistical curvature* of the family at θ_0. Its magnitude is uniquely determined by the above reduction process. Alternately, in an arbitrary curved exponential family it has the formula

(4)
$$\rho^{-1}(t_0) = \left(\frac{|M|}{m_{11}^3}\right)^{\frac{1}{2}}$$

where

$$M = \begin{pmatrix} \dot\theta'\Sigma\dot\theta & \dot\theta'\Sigma\ddot\theta \\ \ddot\theta'\Sigma\dot\theta & \ddot\theta'\Sigma\ddot\theta \end{pmatrix}$$

with $\dot\theta = \dot\theta(t_0)$, $\ddot\theta = \ddot\theta(t_0)$, $\Sigma = \Sigma(\theta(t_0))$. See Efron (1975).

Remark on Notation. The general functional notation $\theta(\cdot)$ was introduced in 3.7(1) as $\theta(x) = \xi^{-1}(x)$. We will continue to use this general notation in contexts not involving specific differentiable subfamilies. In contexts involving differentiable subfamilies the notation $\theta(\cdot)$ will usually refer to a (local) parametrization of the subfamily; if so, this fact will be explicitly noted. Although this means that the very convenient notation $\theta(\cdot)$ can henceforth have either of two meanings we hope there will be no confusion -- simply remember that $\theta(\cdot)$ is defined by 3.7(1) except where explicitly stated otherwise.

3.12 Example

Let Z have exponential density, $f_\lambda(z) = e^{-\lambda z} \chi_{(0,\infty)}(z)$, relative to Lebesgue measure. Let $T > 0$ be a fixed constant. Let Y be the truncated variable $Y = \min(Z, T)$ and $X(y) \in R^2$ be

$$X(y) = \begin{cases} (y, 0) & \text{if} \quad y < T \\ (y, 1) & \text{if} \quad y = T \end{cases} .$$

For $\lambda \in (0, \infty)$ the distribution of X form a standard curved exponential family. The dominating measure ν is composed of linear Lebesgue measure on the line $((0, T) \times 0)$ plus a point mass on $(T, 1)$. The parameter space for this family is

$$(1) \qquad \Theta = \{\theta \in R^2 : \theta_1 = -\lambda, \ \theta_2 = -\ln \lambda, \ \lambda \in (0, \infty)\} ,$$

and

$$(2) \qquad \psi(\theta) = \log [\frac{1}{\theta_1} (e^{\theta_1 T} - 1) + e^{\theta_1 T + \theta_2}] .$$

(The natural parameter space is R^2, since ν has bounded support.) Figure 1 displays both Θ and K on a single plot.

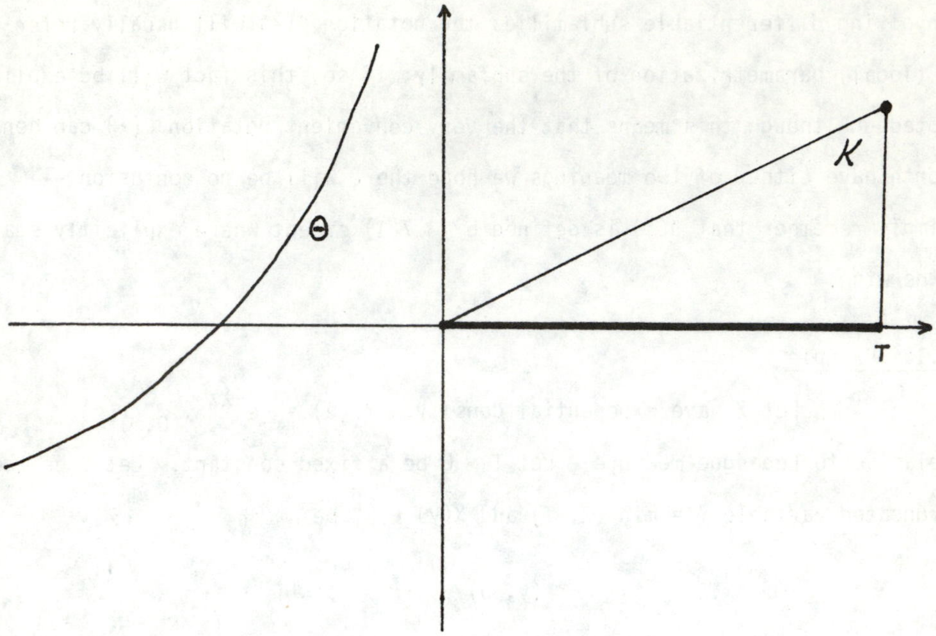

Figure 3.12(1):

Θ and K for Example 3.12.

We return to this example in Chapter 5.

EXERCISES

3.4.1

Let $X_1, X_2, \ldots, X_n$ be a sample from a population with the inverse Gaussian distribution 3.4(1). (i) Show that $S = \sum_{i=1}^{n} X_i$ also has an inverse Gaussian distribution with parameters $\theta_1, n^2\theta_2$. [Examine $E(e^{tS})$.] (ii) Show that S and $(X_i^{-1} - \bar{X}^{-1})$ are independent. [(i) shows that $(S, \frac{n^2}{S}) \sim \text{Expf}(\theta_1, \theta_2)$. Now use Theorem 2.14.]

3.5.1 (Converse to Lemma 3.5)

Let $v \in R^k$, $\alpha \in R$. Let $K \subset N$ be compact. If $\nu(H^{-+}(v, \alpha)) = 0$ then

(1) $$\limsup_{\rho\to\infty} \sup_{\theta \in K} \lambda(\theta + \rho v)/e^{\rho\alpha} = 0 .$$

Also, if $\nu(H^+(v, \alpha)) = 0$ then

(2) $$\limsup_{\rho\to\infty} \sup_{\theta \in K} \lambda(\theta + \rho v)/e^{\rho\alpha} < \infty .$$

(Be careful, these results may be false if $K \not\subset N$.)

In particular, for $\theta \in N$

(3) $$\psi(\theta + \rho v) \to -\infty \quad \text{as} \quad \rho \to \infty$$

if and only if $\nu(\bar{H}^+(v, 0)) = 0$.

3.5.2

Let $Z \in K°$. Let $\varepsilon' = \inf\{||x - Z||: x \notin K\} > 0$. Show

(1) $$\lim_{||\theta||\to\infty} \left(\frac{\psi(\theta) - \theta\cdot Z}{||\theta||}\right) = \varepsilon' .$$

[Translate to the case where Z = 0, using 1.6(3) with $\phi_0 = 0$, $Z_0 = Z$. Then this result is a minor variation of 3.6(3), and could also have been used to establish 3.6(4).]

3.6.1

Is the following assertion a valid converse to Theorem 3.6: Let $\{p_\theta\}$ be a minimal standard exponential family. Then $\xi : N^\circ \to K^\circ$ is a homeomorphism if and only if $\{p_\theta\}$ is steep.(?) [If k = 1 this is easy to prove.]

3.7.1

Define the measure ν on $\{(x_1, x_2) : -\infty < x_1 < \infty\ ,\ x_2 = 0$ or $x_1 = 0,\ x_2 > 0\}$ by

$$\nu((A, 0)) = \int_A c_0 \frac{e^{-|t|}}{1+t^4}\, dt\ , \quad A \subset (-\infty, \infty), \quad \nu((R, 0)) = 1\ ,$$

(1)

$$\nu((0, A)) = \int_A e^{-t}\, dt \qquad A \subset (0, \infty)\ .$$

(i) Show the exponential family generated by ν has $N = \{\theta: -1 \le \theta_1 \le 1,\ \theta_2 < 1\}$ and is *not* steep. (ii) Show that $\xi(N^\circ) \underset{\neq}{\subset} K^\circ = \{x : x_2 \ge 0\}$ and furthermore that $\xi(N^\circ)$ is not even convex. [Show

(2) $$\xi(N^\circ) = \{\xi : |\xi_1| < c\left[1 - \frac{(\xi_2^2 + 4\xi_2 k)^{1/2} - \xi_2}{2k}\right]$$

for appropriate c, k.] See Efron (1978).

3.9.1

Prove the conclusion of Theorem 3.9 if $\{p_\theta\}$ is minimal and steep. [In the proof of Theorem 3.9 let $\phi_2^0 \in N^\circ$ and show (using Definition 3.2) that $\nu_{\phi_2^0}$ is steep. For ease of proof assume (w.l.o.g.) that M = I.]

3.11.1

Verify the formula 3.11(4) for the statistical curvature of a curved exponential family.

3.11.2

(i) Verify 3.10(1). (ii) Reduce the three-dimensional multinomial family to a two-dimensional minimal family and show that 3.10(1) now corresponds to a curved exponential family. (iii) Fix ζ_1 and calculate the statistical curvature of the resulting family as a function of the remaining parameter, ϕ_3. (iv) For what value(s) of ζ_1, ϕ_3 is the curvature zero? Why?

3.11.3

Consider an m-dimensional differentiable subfamily inside a k parameter exponential family. Write a canonical form for this family analogous to that in 4.14(1) - (3). [The case m = 1 required two canonical parameters -- b, ρ -- in 4.14(3). The general case requires $m + m(m + 1)/2$ parameters.]

3.12.1

Let $\{p_\theta\}$ be a canonical k parameter exponential family. Let $\inf\{\psi(\theta): \theta \in N\} < C < \sup\{\psi(\theta): \theta \in N\}$ and let $\Theta = \{\theta \in N^\circ: \psi(\theta) = C\}$. $\{p_\theta: \theta \in \Theta\}$ can be called a *stratum* of $\{p_\theta: \theta \in N\}$. (i) Show that $\{p_\theta: \theta \in \Theta\}$ is a (k - 1) dimensional differentiable subfamily of $\{p_\theta: \theta \in N\}$. (ii) Let $\theta' = (\theta_{(1)}, \theta_{(2)})'$ where $\theta_{(1)}$ is $(k - 1) \times 1$ and $\theta_{(2)}$ is 1×1. Let $\theta(t)$ be any (local) parametrization of $\{p_\theta: \theta \in N\}$ with $t \in T \subset R^{k-1}$. Then

$$(1) \qquad \xi_{(1)}(\theta(t)) \cdot \frac{\partial\theta_{(1)}(t)}{\partial t_j} + \xi_{(2)}(\theta(t)) \frac{\partial\theta_2(t)}{\partial t_j} = 0 \quad .$$

(iii) Let $\theta^\circ \in \Theta$ be any point with $\xi_{(2)}(\theta^\circ) \neq 0$. Then on a neighborhood of θ° in Θ one may write $\theta_{(2)}$ as a function of $\theta_{(1)}$ -- i.e. $\theta_{(2)} = \theta_{(2)}(\theta_{(1)})$ -- and

$$(2) \qquad \nabla\theta_{(2)}(\theta_{(1)}) = -\frac{\xi_{(1)}(\theta)}{\xi_{(2)}(\theta)} \quad .$$

3.12.2

Show that the distributions of X described below can be represented as strata of canonical exponential families (See 3.12.1 for definition.)

(i) $X \sim N(\theta, I)$, $||\theta||^2 = C$.

(ii) The distributions of X-(0,1) with X defined in Example 3.12.

(iii) Let $Y_1, Y_2, \ldots$ be i.i.d. from a canonical regular exponential family, $\{p_\phi\}$. Let N be any Markov stopping time (i.e. $\{y: N(y) = n\}$ is measurable with respect to $Y_1, \ldots, Y_n$). Let $S_n = \sum_{i=1}^{n} Y_i$. Let $X = (S_N, N) = (X_{(1)}, X_{(2)})$, and consider only values of ϕ such that $P_\phi(N < \infty) = 1$. [Let $\theta = (\phi - \psi(\theta))$ where $\psi(\phi)$ is the cumulant generating function for the original family $\{p_\phi\}$.]

3.12.3

In 3.12.2 (iii) show that 3.12.1(2) is identical to the following conclusion also derivable from the martingale stopping theorem:

(1) $$E(S_N) = E(Y)\,E(N) .$$

($(S_n - n\,E(Y)$ is a martingale and so (1) also follows from the stopping theorem applied to this martingale.)

3.12.4

(i) For the family in 3.12.1(i) show the statistical curvature is the constant $1/\sqrt{C}$. (ii) Calculate the statistical curvature for the families described in Example 3.12 and Exercise 3.12.1(ii).

3.12.5

A Poisson process on [0, 1] with intensity function $\rho(t) \geq 0$ may be characterized by the property that the number of observations in any interval $(a, b) \subset [0, 1]$ has $P(\int_a^b \rho(t)dt)$ distribution, and the number of observations in disjoint intervals are independent random variables. Let

$T_1 < \ldots < T_Y$ denote the observations from a Poisson process on [0, 1]. Suppose

(1) $$\rho(t) = \prod_{i=1}^{m} \rho_i^{\alpha_i}(t)$$

where $\rho_i > 0$ are known (measurable) functions on [0, 1] and α_i are unknown parameters. Show that the distributions of $(T_1, \ldots, T_Y, Y)$ form a differentiable subfamily of dimension m in an (m + 1) parameter exponential family. Identify the canonical statistics and observations for this family. Is this family a stratum of the original family? [The conditional distribution of $T_1, \ldots T_Y$ given Y is that of an ordered sample of Y independent observations from a distribution on [0, 1] with density proportional to $\prod_{i=1}^{m} \rho_i^{\alpha_i}(t)$.]

3.12.6

Let Z_{ij} be independent identically distributed variables with a power series distribution:

(1) $$P(Z_{ij} = z) = C(\lambda)\, h(z)\lambda^z, \qquad z=0,1,\ldots, \qquad \lambda > 0.$$

Let $Y_0 = 1$ and define $Y_1, \ldots$ inductively as $Y_i = \sum_{j=1}^{Y_i - 1} Z_{ij}$. $Y_0, Y_1, \ldots$ is called the *Galton-Watson* process with generating distribution (1). Fix $2 \le n < \infty$. Show that the distributions of $Y_0, Y_1, \ldots, Y_n$ form a curved exponential family with natural statistics $(\sum_0^{n-1} Y_j, \sum_0^{n} Y_j)$ and this curved exponential family is a stratum of the corresponding full exponential family.

CHAPTER 4. APPLICATIONS

This chapter describes three different general applications of the theory developed so far. The first part of the chapter contains a proof of the information inequality and a proof based on this inequality of Karlin's theorem on admissibility of linear estimators.

The second part of the chapter describes Stein's unbiased estimate of the risk and proves the minimaxity of the James-Stein estimator as a specific application of this unbiased estimate.

The third part of the chapter describes generalized Bayes estimators and contains two principle theorems describing situations in which all admissible estimators are generalized Bayes -- or at least have a representation similar to that of a generalized Bayes procedure. This part of the chapter deals with two basic situations. The first is estimation of the natural parameter under squared error loss, and the second is estimation of the expectation parameter under squared error loss. The so-called conjugate priors play a natural role in this second situation.

The exercises at the end of the chapter contain a non-systematic selection of some of the specific results derivable from the more general development in the body of the chapter.

INFORMATION INEQUALITY

The information inequality -- also known as the Cramer-Rao inequality -- is an easy consequence of Corollary 2.6.

The version to be proved below applies to vector-valued as well as real-valued statistics. For vector-valued statistics one needs the multivariate Cauchy-Schwarz inequality, as described in the following theorem.

If A,B are symmetric (m×m) matrices, write $A \geq B$ to mean that $A - B$ is positive semi-definite.

<u>4.1 Theorem</u>

Let T_1, T_2 be, respectively (ℓ×1) and (m×1) vector-valued random variables on some probability space. Let

$$B_{11} = E(T_1 T_1') \qquad (\ell \times \ell)$$

$$B_{12} = E(T_1 T_2') \qquad (\ell \times m)$$

$$B_{22} = E(T_2 T_2') \qquad (m \times m)$$

and suppose B_{11} exists and B_{22} exists and is non-singular. Then

(1) $$B_{11} \geq B_{12} B_{22}^{-1} B_{21} \quad .$$

Remarks. If $\ell = m = 1$ this is the usual Cauchy-Schwarz inequality:

(2) $$E(T_1^2)E(T_2^2) \geq E^2(T_1 T_2) \quad .$$

If B_{22} is singular the inequality (1) remains true with generalized inverses in place of true inverses. See Exercise 4.1.1.

If 4.1(1) is applied to the random vectors $T_1 - E(T_1)$, $T_2 - E(T_2)$ it yields the covariance form of the inequality:

(3) $$\Sigma_{11} \geq \Sigma_{12} \Sigma_{22}^{-1} \Sigma_{21} \quad .$$

Proof. Consider the $((\ell + m) \times 1)$ random vector $U = \begin{pmatrix} T_1 \\ T_2 \end{pmatrix}$. Then

$$0 \leq E(U U') = \begin{pmatrix} B_{11} & B_{12} \\ B_{21} & B_{22} \end{pmatrix}$$

Let $W = \begin{pmatrix} I & -B_{12}B_{22}^{-1} \\ 0 & B_{22}^{-1} \end{pmatrix}$. Then

$$0 \leq E(WUU'W') = W E(UU')W'$$

$$= \begin{pmatrix} B_{11} - B_{12}B_{22}^{-1}B_{21} & 0 \\ 0 & B_{22}^{-1} \end{pmatrix} .$$

It follows that $0 \leq B_{11} - B_{12}B_{22}^{-1}B_{21}$, as desired. ||

One further preparatory lemma is needed for the form of the information inequality which appears below.

4.2 Proposition

Let $\{p_\theta\}$ be a standard k-parameter exponential family. Let T be a statistic taking values in R^ℓ. Suppose $\theta_0 \in N^\circ$ and the covariance matrix $\not{Z}_{\theta_0}(T)$ of T exists at θ_0. Then $E_\theta(T)$ exists on a neighborhood of θ_0. ($\theta \in N^\circ(||T||)$ in the notation of 2.6.)

Proof.   For some $\varepsilon > 0$, $||\theta - \theta_0|| < \varepsilon$ implies $\theta \in N$. Let $||\theta - \theta_0|| < \varepsilon/2$. Then, by the ordinary Cauchy-Schwarz inequality,

(1) $E_\theta(||T||) = \int ||T(x)|| \exp(\theta \cdot x - \psi(\theta))\nu(dx)$

$= \int ||T(x)|| \exp((\theta - \theta_0) \cdot x - \psi(\theta) + \psi(\theta_0)) \exp(\theta_0 \cdot x - \psi(\theta_0))\nu(dx)$

$\leq [\int ||T(x)||^2 \exp(\theta_0 \cdot x - \psi(\theta_0)) \nu(dx)$

$\int \exp(2(\theta - \theta_0) \cdot x - 2\psi(\theta) + 2\psi(\theta_0))\exp(\theta_0 \cdot x - \psi(\theta_0))\nu(dx)]^{\frac{1}{2}}$

$= E_{\theta_0}^{\frac{1}{2}} (||T(x)||^2)[\exp \psi(2(\theta - \theta_0) + \theta_0) - 2\psi(\theta) + \psi(\theta_0)]^{\frac{1}{2}}$

$< \infty$

since $E_{\theta_0}(||T(x)||^2) < \infty$ by assumption and since $2(\theta - \theta_0) + \theta_0 \in N$. ||

4.3 Setting

The following version of the information inequality applies to differentiable exponential subfamilies, as defined at the end of Chapter 3.

Let $\{p_\theta: \theta \in \Theta\}$ be such a family with Θ m-dimensional. Let $\theta_0 \in \Theta$. For N a neighborhood in R^m let $\theta : N \to \Theta \subset R^k$, with $\theta(\rho_0) = \theta_0$ be a parametrization of Θ in a neighborhood of θ_0. By definition $\nabla\theta(\rho)$ is the m×k matrix with elements

(1) $$(\nabla\theta(\rho))_{ij} = \frac{\partial\theta_j(\rho)}{\partial\rho_i} \qquad 1 \le i \le m, \qquad 1 \le j \le k \quad .$$

The parametrization can always be chosen so that $\nabla\theta(\rho)$ is of rank m, and we assume this is so.

Define the *information matrix* $J(\rho)$ at $\rho = \rho_0$ by

(2) $$J(\rho_0) = (\nabla\theta(\rho_0))(\Sigma(\theta_0)(\nabla\theta(\rho_0))' \quad .$$

If $\{p_\theta\}$ is a minimal exponential family then $\Sigma(\theta_0)$ is non-singular, and so $J(\rho_0)$ is then a positive definite m×m symmetric matrix. The chain rule and the basic differentiation formula 2.3(2) yield two alternate expressions for J; namely

(3) $$(J(\rho_0))_{ij} = E_{\theta_0}\left(\frac{\partial \log p_{\theta(\rho_0)}(X)}{\partial\rho_i}\;\frac{\partial \log p_{\theta(\rho_0)}(X)}{\partial\rho_j}\right)$$

$$= -E_{\theta_0}\left(\frac{\partial^2 \log p_{\theta(\rho_0)}(X)}{\partial\rho_i\;\partial\rho_j}\right) \quad .$$

The first expression of (3) is, of course, the usual definition of J in contexts more general than differentiable subfamilies.

If T is a statistic taking values in R^ℓ let

(4) $$e(\rho) = e_T(\rho) = E_{\theta(\rho)}(T) \quad .$$

Suppose $\theta_0 \in N^\circ(||T||)$. Then $E_\theta(T)$ and its derivatives exists at θ_0 by Corollary 2.6. The chain rule then yields

(5) $$\nabla e(\rho_0) = (\nabla\theta(\rho_0))(\nabla E_{\theta_0}(T)) \quad .$$

(The preceding formulation of course includes the case where $\{p_\theta\}$ is a full exponential family. Simply set $\rho = \theta$ so that $\theta(\rho) \equiv \theta$. In that case $J(\rho_0) = \Sigma(\theta_0)$ and $\nabla e(\rho_0) = \nabla E_{\theta_0}(T)$.)

4.4 Theorem (Information inequality)

Let $\{p_\theta: \theta \in \Theta\}$ be a differentiable subfamily of a canonical exponential family with $\theta_0 = \theta(\rho_0)$, as above. Let T be an ℓ-dimensional statistic. Suppose $\Sigma_{\theta_0}(T)$ exists. Then $e(\rho) = E_{\theta(\rho)}(T)$ exists and is differentiable on a neighborhood of ρ_0, and the covariance matrix of T satisfies

(1) $$\Sigma_{\theta_0}(T) \geq (\nabla e(\rho_0))' J^{-1}(\rho_0)(\nabla e(\rho_0)) \quad .$$

Proof. $\theta_0 \in N^\circ(||T||)$ by Proposition 4.2. Now apply the Cauchy-Schwarz inequality 4.1(1) with $T_1 = T - E_{\theta_0}(T)$ and

(2) $$T_2(X) = \nabla \ln p_{\theta(\rho_0)}(X) = (\nabla\theta(\rho_0))\,(X - \xi(\theta_0)) \quad .$$

Then $B_{11} = \Sigma_{\theta_0}(T)$,

(3) $$B_{22} = E(T_2\, T_2') = (\nabla\theta(\rho_0))\, \Sigma(\theta_0)(\nabla\theta(\rho_0))' = J(\rho_0) \ ,$$

and

(4) $$B_{12} = E(T_1\, T_2') = (\nabla\theta(\rho_0))(\nabla E_{\theta_0}(T)) = \nabla e(\rho_0)$$

by 2.6(3) and 4.3(5). The Cauchy-Schwarz inequality says $B_{11} \geq B_{12}B_{22}^{-1}\, B_{21}$ which is the same as (1). ||

A useful feature of the form of Theorem 4.4 is the absence of any regularity condition on T other than the existence of $\Sigma_{\theta_0}(T)$. Many other versions of the information inequality contain further assumptions about T (See e.g. Lehmann (1983, Theorem 7.3).) but these are superfluous here.

An information inequality like Theorem 4.4 is needed for applications of the following type.

4.5 Application (Karlin's Theorem on Admissibility of Linear Estimates)

The information inequality can sometimes be used to prove admissibility. In these situations other, more flexible, proofs can also be used, but the information inequality proof is nevertheless easy and revealing. The following result is due to Karlin (1958). The information inequality proof, due to Ping (1964), is a generalization of the first proof of this sort in Hodges and Lehmann (1951). See Lehmann (1983, p.271) for further references and details of the proof.

Theorem. Let $\{p_\theta\}$ be a full regular one-dimensional exponential family with $N = (\underline{\theta}, \bar{\theta})$, $-\infty \leq \underline{\theta} < \bar{\theta} \leq \infty$. Consider the problem of estimating $\xi(\theta) = E_\theta(X)$ under squared error loss. The risk of any (non-randomized) estimator δ is thus $R(\theta, \delta) = E_\theta((\delta(x) - \xi(\theta))^2)$. Then the linear estimator

$$\delta_{\alpha,\beta}(x) = \alpha x + \beta \tag{1}$$

is admissible if $0 < \alpha \leq 1$ and if

$$\int \exp(-\gamma\theta + \lambda\psi(\theta))\, d\theta \tag{2}$$

diverges at both $\underline{\theta}$ and $\bar{\theta}$, where γ, λ are defined by

$$\alpha = \frac{1}{1+\lambda}, \qquad \beta = \frac{\gamma}{1+\lambda} \quad . \tag{3}$$

Proof. We consider here only the case $\beta = 0 = \gamma$. (See Exercise 4.5.1.) Fix α. Let δ be any estimator with finite risk. Let $b(\theta) = E_\theta(\delta(X)) - \alpha\xi(\theta)$. The information inequality yields

$$\begin{aligned} R(\theta, \delta) &\geq \frac{[(\alpha\xi(\theta) + b(\theta))']^2}{\xi'(\theta)} + (\xi(\theta)(1-\alpha) - b(\theta))^2 \\ &\geq \alpha^2\xi'(\theta) + 2\alpha b'(\theta) + (\xi(\theta)(1-\alpha) - b(\theta))^2 \end{aligned} \tag{4}$$

since $\xi(\theta) = E_\theta(X)$ and $\xi'(\theta) = J(\theta) = \mathrm{Var}_\theta X$. For $\delta_{\alpha,0}$

$$R(\theta, \delta_{\alpha,0}) = \alpha^2\xi'(\theta) + (1 - \alpha)^2 \xi^2(\theta) . \tag{5}$$

Hence, if

$$R(\theta, \delta) \leq R(\theta, \delta_{\alpha,0}) \tag{6}$$

then

$$2b'(\theta) - 2\lambda\xi(\theta)\, b(\theta) + (1 + \lambda)\, b^2(\theta) \leq 0 \quad . \tag{7}$$

Let

$$K(\theta) = e^{\lambda\psi(\theta)}\, b(\theta) \quad .$$

Then (7) becomes

$$2K'(\theta) + (1 + \lambda)\, K^2(\theta)e^{\lambda\psi(\theta)} \leq 0 \quad . \tag{8}$$

Now, let $\theta_0 \in (a, b)$ and make the change of variables

$$t(\theta) = \int_{\theta_0}^{\theta} \exp(\lambda\psi(t))dt.$$

Correspondingly, define $k(t)$ by $k(t(\theta)) = K(\theta)$, so that (8) becomes

$$2k'(t) + (1 + \lambda)\, k^2(t) \leq 0 \tag{9}$$

where $-\infty < t < \infty$ by (2). The only solution of (9) for $t \in (-\infty, \infty)$ is $k \equiv 0$ since integration of (9) shows that for $t > t_1$ k is non-increasing and

$$k^{-1}(t) - k^{-1}(t_1) \geq (1 + \lambda)(t - t_1)/2 \quad ;$$

and hence $k(t_1) < 0$ is impossible. A similar inequality for $t < t_1$ shows that $k(t_1) > 0$ is also impossible. It follows that (6) implies $b \equiv 0$, which in turn implies $\delta = \delta_{\alpha,0}$ (a.e.(ν)) by completeness. This proves admissibility of $\delta_{\alpha,0}$. ||

It is generally conjectured that the condition 4.5(2) is necessary as well as sufficient for admissibility of $\delta_{\alpha,\beta}$. However only partial results are known in this connection. See Joshi (1969) and also Exercises 4.5.4, 4.5.5.

4.6 Further Developments

It is useful in considering asymptotic theory to have available a few further results concerning the information inequality. These results are sketched below; the proofs are left for exercises. These results have nothing to do specifically with exponential families but only require a setting in which the information inequality is valid. Nevertheless, for precision assume below the setting of Theorem 4.4, and let $S \subset R^m$ denote a (possibly large) open set on which $\Sigma_{\theta(\rho)}(T)$ exists. For convenience we consider below only estimation of ρ under the quadratic type loss function

$$L(\rho, \delta) = (\delta - \rho)' J(\rho)(\delta - \rho) , \tag{1}$$

and under a truncated version of this loss. (See (3) below.) For proof of the following assertions see Exercises 4.6.1 - 4.6.7 and Brown (1986).

Let h be an absolutely continuous probability density on S, supported on a compact subset $H \subset S$. Then the expected risk satisfies

$$\int_H R(\rho, \delta)h(\rho)d\rho \geq m - \int_H \left(\frac{\nabla h(\rho)}{h(\rho)}\right)' J^{-1}(\rho)\left(\frac{\nabla h(\rho)}{h(\rho)}\right) h(\rho)\, d\rho \quad . \tag{2}$$

Note that the right side of this inequality is independent of δ, and thus provides a lower bound for the Bayes risk under the prior density h.

A natural truncation of the loss (1) is the function $\min(L(\rho, \delta), K)$. Generalizations of the information inequality and of (2), like those to be described below, can be stated for this natural truncation; however the statements and proofs are easier under a different truncation which is equally useful in asymptotics. This truncation will now be described.

Let $K > 0$. For $v \in R$ define

$$v_K = \begin{cases} -K & v < -K \\ |v| & v \le K \\ K & v > K \end{cases} .$$

For $v \in R^k$ define v_K to be the vector with coordinates $(v_K)_i = (v_i)_K$, $i=1,\ldots,k$. Now let

$$L_K(\rho, \delta) = (\delta - \rho)'_K J^{-1}(\rho)(\delta - \rho)_K \quad . \tag{3}$$

Let R_K denote the risk function corresponding to this truncated loss function.

If δ is an estimator of ρ, let

$$\delta_{(K)}(x; \rho) = \rho + (\delta(x) - \rho)_K \quad , \tag{4}$$

and

$$b_{(K)}(\rho) = E_\theta(\delta_{(K)}(X, \rho)) - \rho = e_{(K)}(\rho) - \rho \quad . \tag{5}$$

Let $\lambda_1(\rho) \ge \ldots \ge \lambda_m(\rho) > 0$ denote the ordered eigenvalues of $J(\rho)$. Let α be any number satisfying $0 < \alpha < 1$. Then

$$\left(1 + \frac{\alpha}{(1 - \alpha)\lambda_m K^2}\right) R_K(\rho, \delta) \tag{6}$$

$$\ge \alpha \operatorname{Tr}(J(\rho)(\nabla e_{(K)}(\rho))' J^{-1}(\rho)(\nabla e_{(K)}(\rho))) + \operatorname{Tr}(J(\rho)b_{(K)}(\rho)b'_{(K)}(\rho)) .$$

(Note: $\nabla e_{(K)}$ exists except possibly for a countable number of values of ρ. At these values interpret the right side of (6) as its lim sup; or use right (or left) partial derivatives in place of $\nabla e_{(K)}$, for these always exist.)

This inequality becomes more interesting as K gets large relative to $1/\lambda_m$, for then α can be chosen near 1 but so that $\frac{\alpha}{(1-\alpha)\lambda_m K^2}$ is small.

The inequality (6) leads to an inequality concerning the Bayes risk just as the usual information inequality leads to (2). With h as in (2)

$$(7)\qquad \left(1 + \frac{\alpha}{(1-\alpha)\lambda_m K^2}\right) \int_H R_K(\rho,\ \delta)h(\rho)d\rho$$

$$\geq \alpha m - \alpha^2 \int_H \left(\frac{\nabla h(\rho)}{h(\rho)}\right)' J^{-1}(\rho)\left(\frac{\nabla h(\rho)}{h(\rho)}\right) h(\rho)d\rho \quad .$$

The above bound, unlike (6), does not involve δ (through $e_{(K)}$).

UNBIASED ESTIMATES OF THE RISK

An unbiased estimate of the risk as a tool for proving inadmissibility of estimators first appears in Stein (1973), and has been widely exploited since then. The basic technique is embarassingly simple. It involves merely an integration by parts which succeeds because of the term $e^{\theta\cdot x}$ appearing in the exponential density. Here we describe the method and a few of the easier applications. For further (more complex) applications, see, for example, Berger (1980b), Berger and Haff (1981), and Haff (1983). Here is the heart of the method.

A function $t : R^k \to R$ is called *absolutely continuous* if $t(x_1,\ldots,x_k)$, is absolutely continuous in x_i, $i=1,\ldots,k$, when all x_j, $j\neq i$ are held fixed. Let $t_i' = \frac{\partial t}{\partial x_i}$.

4.7 Theorem

Let $s : R^k \to R$ be absolutely continuous. Assume

(1) $$\int |s(x)|e^{\theta\cdot x}\, dx < \infty \quad , \quad \text{and}$$

(2) $$\int |s_i'(x)|e^{\theta\cdot x}\, dx < \infty, \quad i = 1,\ldots,k.$$

Then

(3) $$\theta_i \int s(x)e^{\theta\cdot x}\, dx = -\int s_i'(x)e^{\theta\cdot x}\, dx \quad .$$

Proof. Set $i = 1$ for convenience. For almost every $(x_2,\ldots,x_k)$

(4) $$\int |s(x_1,\ x_2,\ldots,x_k)|e^{\theta\cdot x}\, dx_1 < \infty$$

and

(5) $$\int |s_1'(x_1,\ x_2,\ldots,x_k)|e^{\theta\cdot x}\, dx_1 < \infty$$

because of (1), (2). For any such $(x_2,\ldots,x_k)$ integration by parts yields

(6) $$\theta_1 \int s(x_1, x_2,\ldots,x_k)e^{\theta\cdot x}\,dx_1$$

$$= \lim_{B\to\infty} \theta_1 \int_{-B}^{B} s(x_1,x_2,\ldots,x_k)e^{\theta\cdot x}\,dx_1$$

$$= \lim_{B\to\infty} \left\{ -\int_{-B}^{B} s_1'(x_1,x_2,\ldots,x_k)e^{\theta\cdot x}dx_1 + \Big[s(x_1,x_2,\ldots,x_k)e^{\theta\cdot x}\Big]_{x_1=-B}^{B}\right\}$$

$$= -\int s_1'(x_1,x_2,\ldots,x_k)e^{\theta\cdot x}dx_1 + \liminf_{B\to\infty} \Big[s(x_1,x_2,\ldots,x_k)e^{\theta\cdot x}\Big]_{x_1=-B}^{B}$$

$$= -\int s_1'(x_1,x_2,\ldots,x_k)e^{\theta\cdot x}\,dx$$

by (2) and then (1). Integration over $x_2,\ldots,x_k$ then yields (3). ||

The assumptions (1) and (2) are slightly more stringent than necessary, and also can be given alternate forms. For example the assumption (5) together with

(7) $$\lim_{x_1\to\pm\infty} s(x_1, x_2,\ldots,x_k)e^{\theta\cdot x} = 0$$

for almost every $x_2,\ldots,x_k$ implies (4), and hence (3) when $i = 1$. Or, for example, when $k = 1$ a potentially useful result is the equality

(8) $$\int_0^\infty \theta s(x)e^{\theta x}\,dx = -\int_0^\infty s'(x)e^{\theta x} - s(0^+)$$

for absolutely continuous functions s having $\int|s'(x)|e^{\theta x}dx < \infty$ and $\lim_{x\to\infty} s(x)e^{\theta x} = 0$. However, the version of the theorem given above suffices for the usual applications.

Theorem 4.6 can be expressed in other forms which are more suggestive of its applications, as in the following two corollaries.

4.8 Corollary

Let $p_\theta(x)$ be a probability density on R^k (relative to Lebesgue measure) of the form

(1) $$p_\theta(x) = h(x) \exp(\theta \cdot x - \psi(\theta))$$

where $h \geq 0$ is absolutely continuous. Let $t : R^k \to R$ be absolutely continuous. Let $t_i' = \frac{\partial t}{\partial x_i}$. Then

(2) $$\theta_i E_\theta(t) = -E_\theta((t_i' + \frac{h_i'}{h} t))$$

provided both expectations in (2) exist.

Let $t : R^k \to R^k$ be absolutely continuous. Then

(3) $$E_\theta(\theta \cdot t) = -E_\theta(\nabla \cdot t + \frac{\nabla h}{h} \cdot t)$$

where $\nabla \cdot t = \sum_{i=1}^{k} \frac{\partial}{\partial x_i} t_i$, provided that

(4) $$E_\theta(|\frac{\partial t_i}{\partial x_i}|) < \infty$$

and

$$E_\theta(| \frac{\nabla h}{h} \cdot t |) < \infty , \qquad i=1,\ldots,k.$$

(In expressions (2), (3), (4) and similar expressions below define $\frac{h_i'}{h} = 0$ if $h = 0$.)

Proof. For (2) note that $\frac{\partial}{\partial x_i}(th) = (t_i' + \frac{h_i'}{h} t)h$ and apply Theorem 4.7. For (3) apply (2) with $i=1,\ldots,k$ and sum. ||

Remarks. Expression (2) immediately yields

(5) $$\theta E_\theta(t) = -E_\theta(\nabla t + t \frac{\nabla h}{h})$$

provided the expectations exist. (3) can also be derived directly from Green's theorem which implies (under suitable conditions) that

(6) $$\int s(x)(\nabla e^{\theta \cdot x})dx = -\int(\nabla \cdot s(x))e^{\theta \cdot x}\, dx \quad .$$

It can also be worthwhile to apply Theorem 4.7 repeatedly, as in the next proposition which is needed for Theorem 4.10.

4.9 Proposition

Let p_θ be as in Corollary 4.8. Assume that h_i' is also absolutely continous, and that

(1) $$E_\theta\left(\left|\frac{h_i'}{h}\right|\right) < \infty \quad ,$$

and

(2) $$E_\theta\left(\left|\frac{h_{ii}''}{h}\right|\right) < \infty , \qquad i=1,\ldots,k$$

(where $h_{ii}'' = \frac{\partial^2}{\partial x_i^2} h$). Then

(3) $$||\theta||^2 = E_\theta\left(\frac{\nabla^2 h}{h}\right)$$

(where $\nabla^2 h = \sum_{i=1}^{k} h_{ii}''$).

Proof. Apply Theorem 4.6 twice for each $i=1,\ldots,k$ and sum over i. ||

Combining the preceding results yields the following unbiased estimator of risk for squared error loss.

4.10 Theorem

Let $\{p_\theta\}$ be an exponential family whose densities are of the form 4.8(1) with h satisfying 4.9(1), (2). Let $\delta: R^k \to R^k$ be any absolutely continuous estimator of θ. Suppose

(1) $$E_\theta(||\delta||^2) < \infty$$

and

(2) $$E_\theta\left(\left|\delta_i' + \frac{h_i'}{h}\delta\right|\right) < \infty , \qquad i=1,\ldots,k \quad .$$

Then

(3) $$E_\theta(||\delta - \theta||^2) = E_\theta\left(||\delta||^2 - 2\left(\nabla \cdot \delta + \frac{\nabla h}{h} \cdot \delta\right) + \frac{\nabla^2 h}{h}\right)$$

Proof. Note that

$$E_\theta(||\delta - \theta||^2) = E_\theta(||\delta||^2 - 2\theta \cdot \delta + ||\theta||^2) \quad .$$

Now use 4.8(3) and 4.9(3) to arrive at (3). ||

Remarks. The left side of (3) is the risk function for squared error loss. As previously, we frequently use the notation $R(\theta, \delta)$ for a risk function when the loss function (here $||\delta - \theta||^2$) is clear from the context. The integrand of the right side of (3) is free of θ; hence this integrand is an unbiased estimate of $R(\theta, \delta)$. For most applications of (3) one actually needs only an unbiased estimate of $R(\theta, \delta_1) - R(\theta, \delta_2)$ where δ_1 and δ_2 are two given estimators. In that case, the term $||\theta||^2$, leading to $\frac{\nabla^2 h}{h}$ in (3), cancels. Assumption 4.9(2) is therefore not needed to arrive at an unbiased estimate of the form

$$(4) \qquad R(\theta, \delta_1) - R(\theta, \delta_2) = E_\theta(||\delta_1||^2 - ||\delta_2||^2 + 2(\nabla \cdot (\delta_1 - \delta_2) + \frac{\nabla h}{h} \cdot (\delta_1 - \delta_2))) \quad .$$

<u>4.11 Application</u> (James-Stein estimator)

The neatest application of Theorem 4.10 is to prove the minimaxity of the James-Stein estimator for a multivariate normal mean. (The original result in James and Stein (1961) uses a different method of proof.) Let X be k-variate normal, $k \geq 3$, with mean $\xi(\theta) = \theta$ and covariance I. Consider the problem of estimating ξ under squared error loss. The usual estimator $\delta_0(x) = x$ is minimax. However, when $k \geq 3$ it is not admissible. Let

$$(1) \qquad \delta(x) = (1 - \frac{r(||x||)}{||x||^2})x$$

where r is absolutely continuous, non-decreasing, and

$$(2) \qquad 0 \leq r(\cdot) \leq 2(k - 2) \quad .$$

Then

(3) $$R(\theta, \delta) \le R(\theta, \delta_0) = k \quad .$$

Strict inequality holds in (3) except when $r \equiv 0$ or when $r \equiv 2(k - 2)$, as can be seen from (5) below.

The normal density is of the form 4.8(1) and $\frac{\nabla h}{h} = -x$. With δ as in (1)

$$||\delta_0||^2 - ||\delta||^2 + 2\frac{\nabla h}{h} \cdot (\delta_0 - \delta) = - \frac{r^2(||x||)}{||x||^2}$$

so that 4.10(4) yields

(4) $$R(\theta, \delta_0) - R(\theta, \delta) = E_\theta\left(2\nabla \cdot \frac{r(||X||)X}{||X||^2} - \frac{r^2(||X||)}{||X||^2}\right) \quad .$$

(It remains to check the regularity conditions needed for 4.10(4), and these will be discussed below.)

Observe that $\nabla \cdot \frac{x}{||x||^2} = \frac{k-2}{||x||^2}$. Hence (4) yields

(5) $$R(\theta, \delta_0) - R(\theta, \delta) = E_\theta\left(\frac{r(||X||)}{||X||^2}(2(k-2) - r(||X||)) + 2\frac{r'(||X||)}{||X||}\right) \quad .$$

The unbiased estimator of the risk which appears on the right of (5) is non-negative because of (2); hence (3) follows. The first estimator of James and Stein was of the form (1) with $r \equiv k - 2$, which is the best possible constant value of r. However, a better estimator (as also noted by James and Stein) is

(6) $$\delta^+(x) = \left(1 - \frac{k-2}{||x||^2}\right)^+ x$$

which corresponds to the choice

$$r(t) = \min(t^2, k-2) \quad .$$

See Exercise 4.11.1. See also Exercises 4.11.5, 4.17.5, and 4.17.6 for generalizations.

(It is also of interest to note that in general if

$\delta_i = \delta_{0i} + \gamma_i$, i=1,...,k, then 4.10(4) yields

(7) $$R(\theta, \delta_0) - R(\theta, \delta) = E_\theta[\sum_{i=1}^{k} 2 \frac{\partial}{\partial x_i} \gamma_i - \gamma_i^2] .$$

The integrand is formally the same as the Cramer-Rao lower bound (in which b(•) replaces γ(•)). See 4.5(7) (with λ = 0) and Exercise 4.5.6. Hence the fact that the inequality

(8) $$\sum_{i=1}^{k} 2 \frac{\partial}{\partial x_i} \gamma_i - \gamma_i^2 \geq 0$$

has a non-trivial solution if and only if $k \geq 3$ leads to the proof of the fact that $\delta_0(x) = x$ is inadmissible if and only if $k \geq 3$.)

The regularity conditions stated in Theorem 4.10 are not always satisfied by an estimator of the form (1). (If, for example, $r(x) \equiv k-2$ then δ is not continuous at $||x|| = 0$.) Justification of (4) therefore requires a supplementary argument: suppose δ is an estimator of the form (1) with a specified r(•). Let δ_ε be the estimator with r(•) replaced by

(9) $$r_\varepsilon(||x||) = \min(||x||^2/\varepsilon , r(||x||)) .$$

Then δ_ε satisfieds the conditions of Theorem 4.10 so that (4) holds for δ_ε. Passing to the limit as $\varepsilon \downarrow 0$ yields that (4) also holds for δ.

There is a very extensive literature concerning the problem of estimating a multivariate normal mean. For an introduction and some references consult Lehmann (1983, Chapter 4).

4.12 Remark

For discrete exponential families there is an analog of the unbiased estimates in 4.8 and 4.10 which involves difference operators instead of partial derivatives. These results are based on the deceptively simple equality

(1) $$\sum_{x=0}^{\infty} \lambda h(x)\lambda^x = \sum_{x=1}^{\infty} h(x-1)\lambda^x$$

They have been particularly useful for certain problems involving Poisson or negative binomial variables. See Hudson (1978), Hwang (1982), and Ghosh, Hwang, and Tsui (1983) for some theory and applications.

GENERALIZED BAYES ESTIMATORS OF CANONICAL PARAMETERS

We first define the concept of a generalized Bayes estimator in the current context and state some foundational results. Then we discuss estimation of the canonical parameter of an exponential family. Later in this chapter we discuss estimation of the expectation parameter, including the topic of conjugate priors for exponential families.

4.13 Definition

Let $\{p_\theta: \theta \in \Theta\}$ be an exponential family of densities. Let $\zeta: \Theta \to R^\ell$ be measurable. Let G be a non-negative (σ-finite) measure on Θ, locally finite at every $\theta \in \Theta$. G is called a *prior measure* on Θ. Let $S \subset R^k$. Then $\delta: S \to R$ is *generalized Bayes* on S (for estimating ζ under squared error loss) if

(1) $$\delta(x) = \frac{\int \zeta(\theta) p_\theta(x) G(d\theta)}{\int p_\theta(x) G(d\theta)}, \qquad x \in S ,$$

where both numerator and denominator exist for all $x \in S$. We say δ is *generalized Bayes* if it is generalized Bayes on S where $\nu(S^c) = 0$. We will use the symbol δ_G to denote the generalized Bayes procedure for G, when this exists.

If the loss is squared error loss --

(2) $$L(\theta, a) = ||a - \zeta(\theta)||^2 \qquad --$$

for estimating $\zeta(\theta)$ and if the Bayes risk,

(3) $$B(G) = \inf_{\delta} B(G, \delta') = \inf_{\delta'} \int R(\theta, \delta') G(d\theta)$$

$$= \inf_{\delta'} \int E_\theta(L(\theta, \delta'(X)) G(d\theta),$$

satisfies $B(G) < \infty$. Then by Fubini's theorem any *Bayes estimator* for G (i.e. one which minimizes $B(G, \delta)$) must also be generalized Bayes for G. One of the topics in which we shall be interested below is that of characterizing complete classes of procedures under squared error loss (2). Since L is strictly convex the nonrandomized procedures are a complete class. The following theorem is our main tool for proving complete class theorems. (In the current context a *complete class* is a set of procedures which contains all admissible procedures.)

4.14 Theorem

With $\{p_\theta\}$ and L as above every admissible procedure must be a limit of Bayes estimators for priors with finite support. More precisely, to every admissible procedure corresponds a sequence G_i of prior distributions supported on a finite set (and hence having finite Bayes risk) such that

$$(1) \qquad \delta_{G_i}(x) \to \delta(x) \quad \text{a.e.}(\nu) \quad ,$$

where (as above) δ_{G_i} denotes the Bayes estimator for G_i.

Proof. This theorem is apparently "well known". Its proof is outside the intended scope of our manuscript. However, I do not know any adequate published reference for it, so a proof is given in the appendix to the monograph. See Theorem A12. Theorems 3.18 and 3.19 of Wald (1950) come close to the above theorem as do some comments in Sacks (1963) and in Le Cam (1955). ||

We now concentrate on estimation of the canonical parameter. In this case generalized Bayes estimators have a particularly convenient form, as described in the next theorem.

4.15 Theorem

Let $\{p_\theta\}$ be a canonical exponential family and let G be a prior measure on Θ for which the generalized Bayes procedure, δ_G for estimating θ

exists. Define the measure H by

(1) $$H(d\theta) = e^{-\psi(\theta)} G(d\theta)$$

and (as usual) let $\lambda_H(x) = \int e^{\theta \cdot x} H(d\theta)$ denote its Laplace transform. Then δ_G satisfies

(2) $$\delta_G(x) = \nabla \ln \lambda_H(x) = \nabla\psi_H(x) , \qquad x \in K^\circ .$$

(If $\nu(\partial K) = 0$ then, of course, (2) completely defines δ_G since $\nu((K^\circ)^{\text{comp}}) = \nu(\partial K) = 0$.)

Proof. By definition the generalized Bayes procedure is

(3) $$\delta_G(x) = \frac{\int \theta e^{\theta \cdot x} H(d\theta)}{\int e^{\theta \cdot x} H(d\theta)} \quad \text{a.e. } (\nu) .$$

By assumption the integrals on the right of (3) exist a.e.(ν); hence $N_H \supset K^\circ$. The denominator exists on N_H, by definition, and by Theorem 2.2, the numerator exists on N_H° and is given by $\nabla\lambda_H(x)$. This proves (2). ||

If δ is only generalized Bayes on $S \subset K$ relative to G one clearly has an analogous representation of δ on S°, namely

(4) $$\delta(x) = \nabla\psi_H(x) , \qquad x \in S^\circ .$$

An interesting special consequence of the above is that if $k = 1$, and $|\delta(x) - x|$ is bounded, and $\lambda\delta(x)$ is generalized Bayes on K° for $0 < \lambda \leq 1$ then $\delta(x) = x + b$. See Meeden (1976).

The foundation for the following major theorem has been laid above and in Section 2.17. The first theorem of this type was proved by J. Sacks (1963) for dimension $k = 1$. Indeed Sacks claimed, but did not prove, validity of the result for arbitrary dimension. Brown (1971) proved the result for arbitrary dimensions when $\{p_\theta\}$ is a normal location family; and that proof was extended to arbitrary exponential families by Berger and Srinivasan

(1978). The proof below follows Brown and Berger-Srinivasan. The proof of Theorem 4.24 is somewhat more like Sacks' original proof.

<u>4.16 Theorem</u>

Let $\{p_\theta\}$ be a canonical k parameter exponential family. Then δ is admissible under squared error loss for estimating θ only if there is a measure H on $\bar{\Theta} \subset \bar{N}$ such that

$$(1) \qquad \delta(x) = \frac{\int \theta e^{\theta \cdot x} H(d\theta)}{\int e^{\theta \cdot x} H(d\theta)} = \nabla\psi_H(x) , \quad \text{for} \quad x \in K^\circ \quad \text{a.e.}(\nu) \quad .$$

Remarks. The expression (1) implicitly includes the condition $N_H \supset K^\circ$, so that both numerator and denominator in (1) are well defined for all $x \in K^\circ$.

If $H(\bar{\Theta} - \Theta) = 0$ so that $\Theta = \bar{\Theta} \subset N$, then one may define

$$(2) \qquad G(d\theta) = e^{\psi(\theta)} H(d\theta)$$

and rewrite (1) as

$$(3) \qquad \delta(x) = \frac{\int \theta p_\theta(x) G(d\theta)}{\int p_\theta(x) G(d\theta)} , \qquad x \in K^\circ \quad .$$

Thus δ is generalized Bayes on K° relative to G. This observation leads to Corollary 4.17 and to further remarks which appear after the corollary.

Proof. Let δ be admissible. By Theorem 4.14 there is a sequence of prior measures G_i, having finite support, such that $\delta_{G_i}(x) \to \delta_G(x)$ a.e.(ν). Let $x_0 \in K^\circ$ such that $\delta_{G_i}(x_0) \to \delta(x_0)$. Since G_i has finite support $\int e^{\theta \cdot x_0 - \psi(\theta)} G_i(d\theta) < \infty$. Let

$$(2) \qquad \tilde{H}_i(d\theta) = e^{-\psi(\theta)} G_i(d\theta) / \int e^{\zeta \cdot x_0 - \psi(\zeta)} G_i(d\zeta) \quad .$$

This is a normalized version of 4.15(1), so, letting $\psi_i = \psi_{\tilde{H}_i}$,

(3) $$\delta_{G_i}(x) = \nabla\psi_i(x) \quad .$$

Since $\int e^{x_0\cdot\theta}\tilde{H}_i(d\theta) = 1$ we assume without loss of generality the existence of a limiting measure H, for which $\tilde{H}_i \to H$ weak*. (Apply 2.16(iv) to the measure $e^{x_0\cdot\theta}\tilde{H}_i$ to get $e^{x_0\cdot\theta}\,\tilde{H}_i \to H^*$, say, and let $H = e^{-x_0\cdot\theta}\,H^*$.) Let $x' \in K°$ such that 4.14(1) holds at x'. Then there is a finite set $S \subset K°$ such that 4.14(1) holds on S and such that B = conhull S satisfies $x_0 \in B°$, $x' \in B°$. Let $x \in S$. Then

(4) $$\psi_i(x) - \psi_i(x_0) = \int_0^1 (x - x_0)\cdot\nabla\psi_i(x_0 + \rho(x - x_0))d\rho$$

$$\leq (x - x_0)\cdot\nabla\psi_i(x) \leq ||x - x_0||\;||\delta_i(x)||$$

by Corollary 2.5. (Note that $\psi_i(x_0) \equiv 0$.) It follows that

(5) $$\limsup_{i\to\infty}\ \sup_{x\in S}\psi_i(x) = \sup_{x\in S}||\delta(x)||\;||x - x_0|| < \infty \quad .$$

This is the principle assumption of Theorem 2.17, which now implies the existence of a subsequence $H_{i'}$ and a limiting measure, which must be H, such that $\psi_i(x) \to \psi_H(x)$, $x \in B°$, and also $\nabla\psi_i(x) \to \nabla\psi_H(x)$, $x \in B°$, by 2.17(5). Since $\nabla\psi_i(x') = \delta_i(x') \to \delta(x')$ we have

(4) $$\delta(x') = \nabla\psi_H(x') \quad .$$

This proves (1) since x' is an arbitrary point of $K°$ satisfying 4.14(1), and since 4.14(1) is satisfied a.e.(ν). ||

<u>4.17 Corollary</u>

Suppose Θ is closed in R^k and

(1) $$\nu(\partial K) = 0 \quad .$$

Then the generalized Bayes procedures form a complete class.

Proof. As noted the admissible procedures are a (minimal) complete class. If δ is admissible then for some prior measure H on $\Theta = \bar{\Theta}$

(2) $$\delta(x) = \frac{\int \theta e^{\theta \cdot x} H(d\theta)}{\int e^{\theta \cdot x} H(d\theta)} \quad \text{a.e.}(\nu)$$

by 4.16(1) and (1), above. Let $G(d\theta) = e^{\psi(\theta)} H(d\theta)$ as in 4.16(2) to get the desired representation,

(3) $$\delta(x) = \frac{\int \theta p_\theta(x) G(d\theta)}{\int p_\theta(x) G(d\theta)} \quad \text{a.e. } (\nu) \quad . \qquad ||$$

Remarks. If ν is dominated by Lebesgue measure then (1) holds since the Lebesgue measure of the boundary of any convex subset of R^k is zero. (To see this note that if C is bounded and convex with $0 \in \text{int}C$ then

$$\partial C = \bigcap_{i=1}^{\infty} [(1 + \tfrac{1}{i})C - (1 - \tfrac{1}{i})C] = \bigcap_{i=1}^{\infty} C_i \text{ , say, where (as usual)}$$

$aC = \{x: \exists y \in C, \ x = ay\}$. See e.g. Rockafeller (1970). Then $\int_{aC} dx = a \int_C dx$ so that $\int_{\partial C} dx = \lim \int_{C_i} dx = \lim(\frac{1}{2i}) \int_C dx = 0$. If C is unbounded apply the result for bounded C to $C \cap \{x: ||x|| < b\}$ and let $b \to \infty$.)

If $\nu(\partial K) \neq 0$ then there are, in general, admissible procedures which are not generalized Bayes. See Exercise 4.17.1. Similarly, if Θ is not closed in R^k there will again be admissible procedures which are not generalized Bayes, even when $\nu(\partial K) = 0$. See Exercise 4.17.2. When $\Theta = N$ and the exponential family is regular then Θ is closed if and only if $N = R^k$. Hence when $\Theta \neq R^k$ one cannot assert that all admissible procedures are generalized Bayes. However, the representation 4.16(1) remains valid. This representation is qualitatively similar to a generalized Bayes representation and is generally as useful as one.

Not all estimators which can be represented in the form 4.17(3) or 4.16(1) are admissible. In fact, many are not. Nevertheless, representations of this form are valuable stepping-off points for general admissibility proofs. See Brown (1971, 1979).

The most conspicuous example of an inadmissible generalized Bayes estimator occurs in the problem of estimating a multivariate normal mean already discussed in 4.11. The usual estimator $\delta(x) = x$ is generalized Bayes, but when $k \geq 3$ it is not admissible. When $k \geq 3$ the positive part James-Stein estimator, defined in 4.11(6), dominates $\delta(x) = x$. However, the positive part James-Stein estimator cannot be generalized Bayes (see Example 2.9); hence is itself inadmissible. So far as I know the problem of finding an (admissible) estimator which dominates 4.11(6) remains open. However, theoretical and numerical evidence indicates that such an estimator cannot have a much smaller risk at any parameter point; hence 4.11(6) remains one of the many reasonable alternatives to $\delta(x) = x$ when $k \geq 3$. (See e.g. Berger (1982).)

GENERALIZED BAYES ESTIMATORS OF EXPECTATION PARAMETERS CONJUGATE PRIORS

The statistical problem of estimating the expectation parameter $\xi(\theta)$, is more often of interest than that considered previously, of estimating the natural parameter. (Of course for normal location families the two problems are identical.) In this case, too, there is a representation theorem for generalized Bayes procedures and a complete class theorem based on a representation similar to that of generalized Bayes. (In some (not fully developed) sense the generalized Bayes representation available here is dual to that in the preceding section -- the differentiation operator is with respect to θ and appears inside the integral sign instead of being with respect to x and appearing outside it.) Both these main results are somewhat more limited than those for estimating θ; but are nevertheless useful.

A new feature of considerable statistical interest appears here. The linear estimators are (generalized) Bayes for the conjugate (generalized) priors. This result is presented first; the conjugate priors are defined in

4.18 and the existence and linearity of their (generalized) Bayes procedures is proved in Theorem 4.19.

4.18 Definition

Prior measures having densities relative to Lebesgue measure of the form

(1) $$g(\theta) = C\, e^{\theta\cdot\gamma-\lambda\psi(\theta)} \qquad \gamma \in R^k, \quad \lambda \geq 0 \qquad ,$$

are called conjugate prior measures. Note that if the prior is of the form (1) then the posterior distribution, calculating formally, has the same general form, with new parameters $\gamma + x$ and $\lambda + 1$. For a sample of size n the parameters become $\gamma + s_n = \gamma + \sum_{i=1}^{n} x_i$ and $\lambda + n$. (Note in (1) that $g = 0$ if $\theta \notin N$ since then $\psi(\theta) = \infty$.)

Arguments resembling those in the following proof show that the conjugate prior measure is finite, and hence can be normalized to be a prior probability distribution if and only if

(2) $$\lambda > 0 \quad \text{and} \quad \gamma/\lambda \in K^{\circ} \quad .$$

See Exercise 4.18.1.

For estimating $\xi(\theta) = E_\theta(X)$, under squared error loss, the Bayes procedures for conjugate priors are linear in x. This fact (often under extraneous regularity conditions) has been known for decades. See, for example, De Groot (1970, Chapter 9) and Raiffa and Schlaiffer (1961). The following precise statement and its converse first appeared in Diaconis and Ylvisaker (1979). (See Exercise 4.19.1 for a statement of the converse.)

4.19 Theorem

Let $\{p_\theta\}$ be a regular canonical exponential family and let $g(\theta)$ be a conjugate prior density as defined by 4.18(1). Then the generalized Bayes procedure for estimating $\xi(\theta)$ exists on the set

(1) $$S = \{x : \delta(x) = \frac{x+\gamma}{\lambda+1} \in K^\circ\}$$

and has the linear form

(2) $$\delta(x) = \frac{x}{\lambda+1} + \frac{\gamma}{\lambda+1} = \alpha x + \beta \quad , \qquad x \in S \quad .$$

Remarks. If $\nu(S^c) = 0$ then δ is generalized Bayes. If $0 \in K$ this always occurs for $\gamma = 0$, $\lambda > 0$. It occurs for $\gamma = 0$, $\lambda = 0$ if (and only if) $\nu(\partial K) = 0$. It can occur for other values of γ,λ as well.

If $x \notin S$ then the generalized Bayes procedure does not exist at x since $\int e^{\theta\cdot x-\psi(\theta)} g(\theta)d\theta = \infty$. See Exercise 4.19.1.

For the relation between the condition that $\nu(S^c) = 0$, so that δ is generalized Bayes, and Karlin's condition, 4.5(2), see Exercise 4.19.2.

Proof. Let $x \in S$. The generalized Bayes procedure at x, if it exists, has the form

(3) $$\delta(x) = \frac{\int (\nabla\psi(\theta)) \exp((x+\gamma)\cdot\theta - (\lambda+1)\psi(\theta))d\theta}{\int \exp((x+\gamma)\cdot\theta - (\lambda+1)\psi(\theta))d\theta}$$

because of the form of g and of p_θ, and because $\xi(\theta) = \nabla\psi(\theta)$ on N and $g(\theta) = 0$ for $\theta \notin N$.

If the integrals in the numerator and denominator of (3) exist then Green's theorem in the form of 4.7(3) yields

(4) $$(x+\gamma)\int \exp((x+\gamma)\cdot\theta - (\lambda+1)\psi(\theta))d\theta$$
$$= (\lambda+1)\int (\nabla\psi(\theta)) \exp((x+\gamma)\cdot\theta - (\lambda+1)\psi(\theta))d\theta \quad .$$

Rearranging terms in (4) yields (2). It remains only to verify that the numerator and denominator of (3) exist.

Let $z = \frac{x+\gamma}{\lambda+1}$. $z \in K^\circ$ since $x \in S$.

Hence

(5) $$\liminf_{\|\theta\|\to\infty} \frac{\psi(\theta) - \theta \cdot z}{\|\theta\|} > 0$$

by 3.5.2(1) (or by 3.6(3) and translation of the origin). It follows that for some $\varepsilon > 0$

(6) $$\exp((x + \gamma) \cdot \theta - (\lambda + 1)\psi(\theta)) = O(e^{-\varepsilon\|\theta\|}) \quad .$$

This proves existence of the integral in the denominator of (3).

Now consider $\xi_1 = \frac{\partial\psi}{\partial\theta_1}$ on N. For simplicity of notation below, let $\xi_1(\theta) = 0$ if $\theta \notin N$. Fix $\theta_2,\ldots,\theta_k$. $\xi_1(\theta_1,\theta_2,\ldots,\theta_k)$ is monotone in θ_1 for $\theta \in N$. Thus for some $q = q(\theta_2,\ldots,\theta_k) \in R$, $\xi_1(\theta_1,\theta_2,\ldots,\theta_k) \le 0$ for $\theta_1 < q$ and $\xi_1(\theta_1,\theta_2,\ldots,\theta_k) \ge 0$ for $\theta_1 > q$. Hence

(7) $$\int |\xi_1(\theta_1,\theta_2,\ldots,\theta_k)| \exp((x+\gamma) \cdot \theta - (\lambda+1)\psi(\theta))d\theta_1$$

$$= \lim_{B\to\infty} \int_{-B}^{q} - \xi_1(\theta_1,\theta_2,\ldots,\theta_k) \exp((x+\gamma) \cdot \theta - (\lambda+1)\psi(\theta))d\theta_1$$

$$+ \lim_{B\to\infty} \int_{q}^{B} \xi_1(\theta_1,\theta_2,\ldots,\theta_k) \exp((x+\gamma) \cdot \theta - (\lambda+1)\psi(\theta))d\theta_1 \,.$$

The function $\exp(-(\lambda+1)\psi(\theta_1,\theta_2,\ldots,\theta_k))$ is absolutely continuous in θ_1 since $\{p_\theta\}$ is regular. (If $\{p_\theta\}$ were not regular there could be a discontinuity at the boundary of N.) Let $\theta_q = (q(\theta_2,\ldots,\theta_k), \theta_2,\ldots,\theta_k)$. Ordinary integration by parts yields

(8) $$\lim_{B\to\infty} \int_{-B}^{q} - \xi_1(\theta_1,\theta_2,\ldots,\theta_k) \exp((x+\gamma) \cdot \theta - (\lambda+1)\psi(\theta))d\theta_1$$

$$= \lim_{B\to\infty} \Big\{-(x_1+\gamma_1) \int_{-B}^{q} \exp((x+\gamma) \cdot \theta - (\lambda+1)\psi(\theta))d\theta_1$$

$$+ [\exp((x+\gamma) \cdot \theta - (\lambda+1)\psi(\theta))]_{\theta_1=-B}^{q}\Big\}$$

$$= -(x_1+\gamma_1) \int_{-\infty}^{q} \exp((x+\gamma) \cdot \theta - (\lambda+1)\psi(\theta))d\theta_1 + \exp((x+\gamma)\cdot\theta_q - (\lambda+1)\psi(\theta_q))$$

by (6). Note that (again by (6))

$$(9) \qquad \exp((x+\gamma)\cdot\theta_q - (\lambda+1)\psi(\theta_q)) = O(\exp(-\varepsilon\sum_{j=2}^{k}\theta_j^2)) \quad .$$

Reasoning similarly for the second integral on the right of (7), integrating both integrals over $\theta_2,\ldots,\theta_k$, and using (9) yields

$$(10) \qquad \int_{R^k} |\xi_1(\theta)| \exp((x+\gamma)\cdot\theta - (\lambda+1)\psi(\theta))d\theta < \infty \quad .$$

Finally, the identical reasoning on ξ_i, $i=1,2,\ldots,k$, shows that

$$\int (||\nabla\psi(\theta)||) \exp((x+\gamma)\cdot\theta - (\lambda+1)\psi(\theta))d\theta < \infty \quad ,$$

which verifies that the numerator of (3) exists. As noted previously, this completes the proof. ||

4.20 Application

For a given k-parameter exponential family $\{p_\theta\}$ the conjugate prior distributions, $\{g_{\gamma,\lambda}\}$, say, form a (k+1)-parameter exponential family with canonical statistics $\theta_1,\ldots,\theta_k$, $-\psi(\theta)$. This (k+1)-parameter family is minimal except when $\psi(\theta)$ is a linear function of θ. This linearity occurs when p_θ is the $\Gamma(\alpha, \sigma)$ family with known σ, and in certain multivariate generalizations of this univariate example.

Many familiar exponential families are the conjugate families of prior distributions for other familiar exponential families of distributions. (Conjugate prior measures which are not finite then appear as limits of these distributions.) For example, the $N(\gamma, \lambda^{-1}I)$ distributions are conjugate to the $N(\mu, I)$ family. The proper conjugate prior distributions for the $\Gamma(\alpha, \frac{1}{(-\theta)})$ family (α known, $\theta < 0$) are those of $-\Theta$ where $\Theta \sim \Gamma(\lambda\alpha, -\gamma)$, $\gamma < 0$, $\lambda > 0$. The proper conjugate priors for the $P(e^\theta)$ family have density

$$(1) \qquad g_{\gamma,\lambda}(\theta) = e^{\gamma\theta-\lambda e^\theta}, \qquad \gamma < 0, \qquad \lambda \geq 0$$

with respect to Lebesgue measure on $(-\infty, \infty)$. Thus the density of $\xi = e^\theta$ is

$\Gamma(-\gamma, 1/\lambda)$. See also Exercise 5.6.3.

The basic representation theorem for generalized Bayes procedures is a simple consequence of Green's Theorem 4.7(3), and is an obvious extension of 4.19(4) in the proof of Theorem 4.19. The regularity conditions in the following statement may be modified as noted in the remark following the theorem.

<u>4.21 Theorem</u>

Let $\{p_\theta\}$ be a regular canonical exponential family and let G be a prior measure on Θ. Suppose G has a density, g, with respect to Lebesgue measure. Suppose $g(\theta)e^{-\psi(\theta)}$ is absolutely continuous on R^k. Assume for $x \in S$

(1) $$\int e^{\theta\cdot x-\psi(\theta)}\, g(\theta)d\theta \;<\; \infty \;,$$

(2) $$\int \|\nabla g(\theta)\|\, e^{\theta\cdot x-\psi(\theta)}\, d\theta \;<\; \infty \;,$$

and

(3) $$\int \|\nabla\psi(\theta)\|\, g(\theta)e^{\theta\cdot x-\psi(\theta)}\, d\theta \;<\; \infty \,.$$

Then the generalized Bayes procedure, δ, for estimating $\xi(\theta)$ under squared error loss, exists on S and is given by the formula

$$\delta(x) \;=\; x + \frac{\int(\nabla g(\theta))e^{\theta\cdot x-\psi(\theta)}\, d\theta}{\int g(\theta)e^{\theta\cdot x-\psi(\theta)}\, d\theta} \,.$$

Remarks. If $\nu(S^c) = 0$ then, of course, the unrestricted generalized Bayes procedure exists and is given by (4).

Conditions (1) and (2) are of course necessary for the representation (4) to make sense. Condition (3) is necessary in order that the generalized Bayes estimator be well defined. However it can often be deduced as a consequence of (2) and so then need not be checked directly. Suppose

(5) $$g(\theta)e^{\theta\cdot x-\psi(\theta)} \;\le\; h(\|\theta\|)$$

for some function $h(\theta)$ satisfying

(5')
$$\int_0^\infty t^{k-1}\, h(t)dt \; < \; \infty \; .$$

Then (1) is satisfied, and condition (2) implies condition (3). See Exercise 4.21.1.

The representation (4) is exploited in Brown and Hwang (1982) as the starting point for a proof of admissibility of generalized Bayes estimators under certain (important) extra regularity conditions.

Proof. Conditions (1), (2), and (3) justify use of the integration by parts formula 4.7(3), which yields

(6)
$$\int x(g(\theta)e^{-\psi(\theta)})e^{\theta\cdot x}\, d\theta \; = \; \int (-\nabla g(\theta) + g(\theta)\nabla\psi(\theta))e^{\theta\cdot x-\psi(\theta)}\, d\theta \quad .$$

Rearranging terms (each of which exists by (1), (2), (3)) yields (4). ||

We now turn to the complete class theorem comparable to Theorem 4.16. The result proved below applies only to one parameter exponential families. It appears to us that there exists a satisfactory multiparameter analog of this result which, however, is somewhat more complex to state (and to prove). We hope to present this multiparameter extension in a future manuscript.

As with Theorem 4.16 the representation of admissible procedures involves a ratio of integral expressions similar to the formula for a generalized Bayes estimator. Again, under certain additional conditions, this representation reduces to precisely that of a generalized Bayes procedure. A new complication appears in the integral representation below. It applies only on an interval I_δ whose definition involves $\delta(\cdot)$ itself. (See 4.24(1).) However, as explained in the remarks following the theorem, the values of $\delta(x)$ for $x \notin \bar{I}$ are uniquely specified by monotonicity considerations. Hence the theorem actually describes exactly the values of $\delta(x)$ except for at most two points -- the endpoints of I_δ. In this sense the complication presented by the presence of I_δ is just a minor nuisance.

We begin with a technical lemma.

4.22 Lemma

Let ν_n be a sequence of probability measures on R^1. Suppose for some $\zeta > 0$

(1) $$\liminf_{n\to\infty} \nu_n(\{x > K\}) > \zeta > 0$$

for all $K < \infty$. Let $\varepsilon > 0$. Suppose $\lambda_{\nu_n}(\varepsilon) < \infty$, $n=1,\ldots$. Then

(2) $$\lim_{n\to\infty} \frac{\int_K^\infty e^{\varepsilon x}\,\nu_n(dx)}{\lambda_{\nu_n}(\varepsilon)} = 1$$

for all $K < \infty$.

Remarks. The negation of (1) is the condition

(3) $$\lim_{K\to\infty} \liminf_{n\to\infty} \nu_n(\{|x| > K\}) = 0 \quad .$$

This is the usual necessary and sufficient condition for there to exist a subsequence n' and a non-zero limiting measure ν such that $\nu_{n'} \to \nu$.

The conclusion (2) can be paraphrased by saying that the sequence of probability measures $e^{\varepsilon x}\,\nu_n(dx)/\lambda_{\nu_n}(\varepsilon)$ sends all its mass out to $+\infty$.

Proof. Let $K < \infty$, $1 < m < \infty$. Then

(4) $$\frac{\int_K^\infty e^{\varepsilon x}\,\nu_n(dx)}{\int_{-\infty}^K e^{\varepsilon x}\,\nu_n(dx)} \geq \frac{\int_{mK}^\infty e^{\varepsilon x}\,\nu_n(dx)}{\int_{-\infty}^K e^{\varepsilon x}\,\nu_n(dx)}$$

$$\geq e^{\varepsilon(m-1)K}\,\nu_n(\{x > mK\}) \quad .$$

Now let $n \to \infty$ and $m \to \infty$ to find

$$\liminf_{n\to\infty} \frac{\int_K^\infty e^{\varepsilon x}\, \nu_n(dx)}{\int_{-\infty}^K e^{\varepsilon x}\, \nu_n(dx)} = \infty \qquad (5)$$

which proves (2). ||

4.23 Theorem

Let $\{p_\theta : \theta \in \Theta\}$ be a regular exponential family on R^1. Consider the problem of estimating the expectation parameter, $\xi(\theta)$, under squared error loss. Let δ be an admissible estimator. Then, $\delta(\cdot)$ must be a non-decreasing function. Let

$$I_\delta = \{x: \nu(\{y: y > x, \delta(y) \in K^\circ\}) > 0 \text{ and } \nu(\{y: y < x, \delta(y) \in K^\circ\}) > 0\}. \qquad (1)$$

Then there exists a finite measure V on $\bar{\Theta}$ such that for all $x \in I$

$$\delta(x) = \frac{\int \frac{\xi(\theta)}{1 + |\xi(\theta)|} e^{\theta x}\, V(d\theta)}{\int \frac{1}{1 + |\xi(\theta)|} e^{\theta x}\, V(d\theta)} . \qquad (2)$$

Remarks. In (2) the functions $\frac{\xi(\theta)}{1 + |\xi(\theta)|}$ and $\frac{1}{1 + |\xi(\theta)|}$ have the obvious interpretation on the boundary of N. (In other words, if $N = (a, b)$ then $\frac{\xi(b)}{1 + |\xi(b)|} = 1$, $\frac{\xi(a)}{1 + |\xi(a)|} = -1$, etc., since $\lim_{\theta\uparrow b} \xi(\theta) = \infty$, $\lim_{\theta\downarrow a} \xi(\theta) = -\infty$.)

By monotonicity of δ, I must be an open interval. Say $I = (\underline{i}, \bar{i})$, $-\infty \le \underline{i} < \bar{i} \le \infty$. Suppose $K^\circ = (\underline{k}, \bar{k})$, $-\infty \le \underline{k} < \bar{k} \le \infty$. Then $\underline{k} \le \underline{i}$ ($\bar{i} \le \bar{k}$, respectively) and, by monotonicity and the definition of I, $\delta(x) = \underline{k}$ for $\underline{k} \le x < \underline{i}$ ($\delta(x) = \bar{k}$ for $\bar{i} < x \le \bar{k}$). For $\underline{i} < x < \bar{i}$, $\delta(x)$ is defined by (2). Thus, the theorem fails to define $\delta(x)$ only for $x = \underline{i}$ if $-\infty < \underline{k} < \underline{i}$ or for $x = \underline{k}$ if $-\infty < \underline{k} = \underline{i}$, and, if $\bar{k} < \infty$, for $x = \bar{i}$ or $\bar{k}$ depending on whether $\bar{i} < \bar{k}$ or $\bar{i} = \bar{k}$. If ν, the dominating measure for $\{p_\theta\}$, is continuous then these two points have measure 0 and the theorem completely describes δ. Similarly, if $K = (-\infty, \infty)$ then irrespective of ν the theorem completely describes δ

If $V(\bar{N} - N) = V(\{a, b\}) = 0$ then (2) can be rewritten as

$$(3) \qquad \delta(x) = \frac{\int \xi(\theta)\, p_\theta(x) G(d\theta)}{\int p_\theta(x) G(d\theta)}, \qquad x \in I^\circ ,$$

where

$$G(d\theta) = \frac{e^{\psi(\theta)}}{1 + |\xi(\theta)|} V(d\theta) .$$

Thus, δ is then generalized Bayes on I in the ordinary sense. (This must, of course, occur if $N = R^1$.) When $N \neq R^1$ there may exist admissible procedures having representation (2) but not (3). See Exercise 4.24

Finally, note as with Theorem 4.16 that there are many inadmissible procedures satisfying (2). See for example Exercise 4.5.4.

Proof. If G is a prior density then the Bayes procedure (assuming it is well defined for $x \in K$) is given by the formula

$$(4) \qquad \delta_G(x) = \frac{\int \xi(\theta) e^{\theta x - \psi(\theta)}\, G(d\theta)}{\int e^{\theta x - \psi(\theta)}\, G(d\theta)} = \frac{\int \xi(\theta) e^{\theta x}\, H(d\theta)}{\int e^{\theta x}\, H(d\theta)}$$

where $H(d\theta) = ce^{-\psi(\theta)}\, G(d\theta)$. $\xi(\theta)$ is monotone on N. The family of densities $e^{\theta x}/\int e^{\theta x} H(d\theta)$ is an exponential family (with parameter x) relative to the dominating measure H. In particular, it has monotone likelihood ratio. Hence, δ_G is monotone non-decreasing by Corollary 2.22. (δ_G is actually strictly increasing unless G is concentrated on a single point.) All admissible procedures are (a.e.(ν)) limits of Bayes procedures by Theorem 4.14, and limits of monotone functions are monotone. Hence all admissible procedures must be monotone non-decreasing. (A different proof of a better result is contained in Brown, Cohen and Strawderman (1976).)

Let δ be admissible and let δ_{G_i} be the sequence promised in Theorem 4.14 having $\delta_{G_n} \to \delta$ a.e.(ν). Since all δ_{G_n} are monotone non-decreasing there is no loss of generality in assuming $\delta_{G_n}(x) \to \delta(x)$ for all $x \in K°$, and we do so below.

Assume $0 \in I \subset K°$. Define the probability measures,

$$V_n(d\theta) = \frac{(1 + |\xi(\theta)|)\, e^{-\psi(\theta)}\, G_n(d\theta)}{\int (1 + |\xi(\theta)|)\, e^{-\psi(\theta)}\, G_n(d\theta)} \quad . \tag{5}$$

Let $\varepsilon > 0$ such that $\varepsilon \in K°$. Then

$$\delta_{G_n}(\varepsilon) = \frac{\int \frac{\xi(\theta)}{1 + |\xi(\theta)|} e^{\varepsilon\theta} V_n(d\theta)}{\int \frac{1}{1 + |\xi(\theta)|} e^{\varepsilon\theta}\, V_n(d\theta)} \quad . \tag{6}$$

Suppose for some $\zeta > 0$

$$\liminf_{n\to\infty} V_n(\{\theta > K\}) > \zeta > 0 \qquad \text{for all} \quad K < \infty \quad . \tag{7}$$

Let θ_0 be the unique value such that $\xi(\theta_0) = 0$, and let $K > \theta_0$. The function $\frac{\xi(\theta)}{1 + |\xi(\theta)|}$ is increasing for $\theta > \theta_0$. Apply Lemma 4.23 to get

$$\frac{\int \frac{\xi(\theta)}{1 + |\xi(\theta)|} e^{\varepsilon\,\theta} V_n(d\theta)}{\int e^{\varepsilon\theta}\, V_n(d\theta)} \geq \frac{- \int_{-\infty}^{K} e^{\varepsilon\theta}\, V_n(d\theta) + \frac{\xi(K)}{1 + \xi(K)} \int_K^{\infty} e^{\varepsilon\theta} V_n(d\theta)}{\int e^{\varepsilon\theta}\, V_n(d\theta)} \tag{8}$$

$$\to \frac{\xi(K)}{1 + \xi(K)} \quad .$$

Similarly, $\frac{1}{1 + |\xi(\theta)|}$ is decreasing for $\theta > \theta_0$ so that

$$\frac{\int \frac{1}{1 + |\xi(\theta)|}\, e^{\varepsilon\theta}\, V_n(d\theta)}{\int e^{\varepsilon\theta}\, V_n(d\theta)} \leq \frac{\int_{-\infty}^{K} e^{\varepsilon\theta}\, V_n(d\theta) + \frac{1}{1 + \xi(K)} \int_K^{\infty} e^{\varepsilon\theta}\, V_n(d\theta)}{\int e^{\varepsilon\theta}\, V_n(d\theta)} \tag{9}$$

$$\to \frac{1}{1 + \xi(K)} \quad .$$

Substitute (8) and (9) into the formula, (6), for $\delta_{G_n}(\varepsilon)$ and let $K \to \infty$ to find

(10) $$\delta(\varepsilon) = \lim_{n\to\infty} \delta_{G_n}(\varepsilon) \geq \lim_{K\to\infty} \xi(K) .$$

This holds for all $\varepsilon > 0$ with $\varepsilon \in K^\circ$. It follows from (10) that $0 \notin I$, contrary to assumption. Hence (7) must be false. A symmetric argument shows that $\liminf_{n\to\infty} V_n(\{\theta < -K\}) > \zeta > 0$ is also impossible. Hence

(11) $$\lim_{K\to\infty} \liminf_{n\to\infty} V_n(\{|\theta| > K\}) = 0 .$$

By translating the origin the same argument can be applied at any $x \in I \subset K^\circ$. The conclusion is that $x \in I$ implies

(12) $$\lim_{K\to\infty} \liminf_{n\to\infty} \frac{\int_{|\theta|>K} e^{\theta x} V_n(d\theta)}{\int e^{\theta x} V_n(d\theta)} = 0 .$$

This is slightly more than is needed to apply the corollary of Theorem 2.17 stated in Exercise 2.17.2. ((12) implies 2.17.2(2) with the roles of θ and x interchanged so that $P_{n,x}(d\theta) = e^{\theta x}V_n(d\theta)/ e^{\theta x}V_n(d\theta)$.) The conclusion of this exercise is that there exists a subsequence $\{n'\}$ and a limiting measure V on $\bar{\Theta}$ such that

(13) $$e^{\theta x}V_{n'}(d\theta) \to e^{\theta x}V(d\theta) \quad \text{and} \quad \lambda_{V_{n'}}(x) \to \lambda_V(x) , \quad x \in I^\circ .$$

Note that $V(R^k) = \lambda_V(0) = \lim \lambda_{V_{n'}}(0) = 1$. Since both $\frac{\xi(\theta)}{1 + |\xi(\theta)|}$ and $\frac{1}{1 + |\xi(\theta)|}$ are bounded continuous functions on $\bar{\Theta}$, (13) and (6) yield directly that for $x \in I$

$$\delta(x) = \lim_{n'\to\infty} \delta_{G_{n'}}(x) = \frac{\frac{\xi(\theta)}{1 + |\xi(\theta)|} e^{x\theta} V(d\theta)}{\frac{1}{1 + |\xi(\theta)|} e^{x\theta} V(d\theta)} .$$

This verifies (2), and completes the proof. ||

EXERCISES

4.1.1

(i) Prove the Cauchy-Schwarz inequality (4.1(1)) with B_{22}^- in place of B_{22}^{-1} when B_{22} is singular. (ii) Show the inequality remains valid when T_1, T_2 are respectively $(\ell \times s)$ and $(m \times s)$ matrix valued random variables. [(i) Reproduce the proof of Theorem 4.1 with B_{22}^- in place of B_{22}^{-1}; or rotate coordinates so that B_{22} is diagonal with diagonal entries $d_{ii} > 0$, $1 \le i \le r$, and $d_{ii} = 0$, $r+1 \le i \le m$, and apply 4.1(1) for the first r coordinates of T_2.]

4.2.1

Let ν be a measure on R^k and let T be a real valued statistic. Suppose $0 \in N^\circ$ and $E(T^2) < \infty$. Show for every $\varepsilon > 0$ there is a polynomial p(x) such that $E((T - p)^2) < \varepsilon$. (In other words, the monomials $x_1, \ldots, x_k, x_1^2, x_1 x_2, \ldots$ form a complete basis for $L_2(\nu)$.) [For k = 1 (for simplicity) let $f_0 = 1$, $f_1 = a_{11}x + a_{10}$, $f_2 = a_{22}x^2 + a_{21}x + a_{20}, \ldots$ be orthonormal functions in $L_2(\nu)$. Let $\alpha_i = E(Tf_i)$. Then $\Sigma\alpha_i^2 \le ET^2$ so that $g = \sum_{i=0}^{\infty} \alpha_i f_i \in L_2(\nu)$. $T - g \in L_2(\nu)$, $0 \in N^\circ(T - g)$ and $\lambda_{T-g}^{(j)}(0) = 0$, $j=0,1,\ldots$.]

4.3.1

Verify formulae 4.3(3) and 4.3(5).

4.3.2

Let p_θ be a full canonical exponential family and let $\xi = \xi(\theta)$ denote the expectation parameter. Show that relative to this parameter the information matrix is $J(\xi) = \Sigma^{-1}(\theta(\xi))$.

4.4.1

Let M be a fixed $\ell \times \ell$ positive semi-definite symmetric matrix. Write the information inequality for $E_{\theta_0}((T - \mu)' M(T - \mu))$ where T is an ℓ-dimensional statistic with mean μ and finite covariance at θ_0. [This is immediate from Theorem 4.4 and $(T - \mu)' M(T - \mu) = \mathrm{Tr}(M(T - \mu)(T - \mu)')$.]

4.4.2

Show that the information inequality 4.4(1) is an equality if and only if for some matrix A and vector b

(a) $$T(x) = A(\nabla\theta(\rho_0))X + b \quad .$$

[Show the Cauchy-Schwarz inequality is an equality if and only if T_2 is an affine transformation of T_1.]

4.4.3

Let $\{p_\theta: \theta \in \Theta\}$ be a differentiable subfamily and T an ℓ-dimensional statistic. Suppose $\mathcal{I}_{\theta_0}(T)$ exists for some $\theta_0 \in \Theta$. Then the information inequality is an equality for all $\theta \in \Theta$ if and only if Θ is an affine subspace of N and T is an affine function of the canonical minimal sufficient statistic for the exponential family $\{p_\theta: \theta \in \Theta\}$. (That such a characterization holds under mild regularity conditions for a general family $\{p_\theta\}$ was proved in Wijsman (1973) and Joshi (1976).) [Use Exercise 4.4.2.]

4.4.4

Suppose $\{p_\theta\}$ is a canonical one-parameter exponential family. Show that when the information inequality is not an equality it can be improved to an inequality of the form:

(1) $$\text{Var}_{\theta_0} T \geq \varepsilon'(\theta_0)M(\theta_0)\varepsilon(\theta_0)$$

where $\varepsilon(\theta)$ is the $j\times 1$ vector with

(2) $$\varepsilon(\theta)_i = \frac{\partial^i}{\partial\theta^i} e(\theta) \qquad i=1,\ldots,j$$

and $M(\theta)$ is an appropriate $j\times j$ symmetric matrix, not depending on T. In fact, $M^{-1}(\theta)$ is the covariance matrix at θ of the vector with coordinates

(3) $$p_\theta^{-1} \frac{\partial^i}{\partial\theta_i} p_\theta \qquad i=1,\ldots,j \quad .$$

(The inequality (1) with M^{-1} as in (3) is called a *Bhattacharya* inequality.

Such inequalities are valid also for full k parameter exponential families and for ℓ-dimensional statistics, as well as for differentiable subfamilies (ρ replaces θ in (1) - (3)). See e.g. Lehmann (1983, p.129). [A direct proof is possible which also yields the formula (3). An alternate proof assumes $\theta_0 = 0$, $\psi(\theta_0) = 1$ (w.l.o.g.) and uses Exercise 4.2.1 to write

$$\int (T(x) - \alpha_0)^2 \nu(dx) \geq \sum_{i=1}^{j} \alpha_i^2 = \sum_{i=1}^{j} \int T(x) f_i(x) \nu(dx). \quad]$$

4.4.5

Suppose $X_1, \ldots$ are i.i.d. observations from a differentiable exponential subfamily. Let N be a stopping time with $P_{\theta_0}(N < \infty) = 1$ and

(1) $$E_{\theta_0}(\exp(\varepsilon N)) < \infty \quad \text{for some} \quad \varepsilon > 0 \quad .$$

Let $S_n = \sum_{i=1}^{n} X_i$ and let $T(S_N, N)$ be a statistic for which $\mathcal{V}_{\theta_0}(T) < \infty$. Then

(2) $$\mathcal{V}_{\theta_0}(T) \geq (E_{\theta_0}(N))^{-1} (\nabla e(\rho_0))' J^{-1}(\rho_0)(\nabla e(\rho_0))$$

where $e(\rho) = E_{\theta(\rho)}(T(S_N, N))$. [Prove directly or use Exercise 3.12.2 (iii) and Theorem 4.4. The regularity condition (1) can be considerably relaxed or modified, but some condition on N is needed in general. See Simons (1980).]

4.4.6

(i) When $\{p_\theta\}$ is a full canonical exponential family and $E_{\theta_0}(T^2) < \infty$, the Bhattacharya inequalities 4.4.4(1) tend to equality in the limit as $j \to \infty$. (ii) If $\{p_\theta\}$ is an m-dimensional differentiable subfamily with $m < k$ then there are statistics T for which the appropriate Bhattacharya inequalities do not tend to equality as $j \to \infty$. [(i) Use Exercise 4.2.1 and proceed from the proof sketched in the hint in Exercise 4.4.4. (ii) Consider a curved exponential family in the canonical version 3.11(1), and let $T(x) = x_2 - x_1^2$.]

4.5.1

Prove the assertion in 4.5 when $\beta \neq 0$. [Let $Y = X - \gamma$. Apply 4.5

to yield αY as an admissible estimator of $\xi(\theta) - \gamma$. Hence $\alpha Y + \gamma$ is admissible for $\xi(\theta)$.]

4.5.2

Show the condition 4.5(2) implies $\delta_{\alpha,\beta}(x) = (\alpha x + \beta) \in K$ a.e.(ν). [The theorem would be false otherwise! But a direct proof not involving the theorem is also of interest. Use Lemma 3.5.]

4.5.3

Suppose (λ, γ) satisfies condition 4.5(2), $\lambda' < \lambda$, and either $\gamma \in K^\circ$ or ν is a discrete measure. Then (λ', γ) satisfies condition 4.5(2). If $\gamma \in \partial K = K - K^\circ$, and (λ_1, γ), (λ_2, γ) both satisfy 4.5(2), and $\lambda_1 < \lambda < \lambda_2$ then (λ, γ) satisfies 4.5(2).

4.5.4

Let $X \sim \Gamma(a, \sigma)$, a known, and consider the problem of estimating $\sigma = E(X)$ under squared error loss. (i) Using Karlin's theorem verify that $\delta_{\alpha,\beta}(x) = \alpha x + \beta$ is admissible if $\alpha = \frac{1}{a+1}$, $\beta = 0$ or if $\alpha \leq \frac{1}{a+1}$, $\beta > 0$. (ii) Show that if α, β do not satisfy these conditions then $\delta_{\alpha,\beta}$ is inadmissible since there is an admissible linear estimator which is better.

4.5.5

Consider the one-parameter exponential family defined by 3.4(1) with $\theta_2 = -1$ and $\theta = \theta_1 \in (-\infty, 0)$. Consider the problem of estimating $\xi(\theta)$ under squared error loss. Let $\delta_{\alpha,\beta}$ be a linear estimator as in 4.5(1). Observe that condition 4.5(2) of Karlin's theorem is not satisfied at $\bar{\theta} = 0$. Show that $\delta_{\alpha,\beta}$ is inadmissible. [For the case $\alpha = 1$, $\beta = 0$ let

(1) $$\delta'_c(x) = \begin{cases} x & x \leq c \\ c + (x-c)/2 & x > c \end{cases} .$$

Then $R(\theta, \delta'_c) \leq R(\theta, \delta_{1,0})$ for $\xi(\theta) \leq c$ and, for $\xi(\theta) \geq c$, a crude bound yields

$$(2) \qquad R(\theta, \delta') \le (\tfrac{1}{4}) \operatorname{Var}_\theta(X) + (\tfrac{1}{4})(\xi(\theta) - c)^2 + \xi^2(\theta)$$

$$= \xi^3(\theta)/8 + (\xi(\theta) - c)^2/4 + \xi^2(\theta) \quad .$$

Hence for c sufficiently large $R(\theta, \delta') < R(\theta, \delta_{0,1}) = \xi^3(\theta)/2$ also when $\xi(\theta) \ge c$.]

4.5.6

Let $\{p_\theta\}$ be as in 4.5. Suppose it is desired to estimate $g(\theta) = \xi(\theta) + W'(\theta)$ under squared error loss. Show the estimator $\delta_{\alpha,\beta}$ is admissible if

$$(1) \qquad \int \exp(\lambda\psi(\theta) + (1 + \lambda)W(\theta) - \gamma\lambda(\theta)d\theta$$

diverges at both $\underline{\theta}$ and $\bar{\theta}$. [Define $b(\cdot)$ as in 4.5. 4.5(7) becomes

$$(2) \qquad 2b'(\theta) - 2(\lambda\xi(\theta) + (1 + \lambda)W'(\theta))b(\theta) + (1 + \lambda)b^2(\theta) \le 0 \quad .]$$

(See Ghosh and Meeden (1977). Although an estimator $\delta_{\alpha,\beta}$ may be admissible here, it is not clear that it is desirable, whereas for the case $W \equiv 0$ of 4.5 these estimators are very natural.)

4.5.7

Let $\{p_\theta\}$ be a canonical two dimensional exponential family with $N = R^2$. Consider the problem of estimating $\xi(\theta)$ with squared error loss (so that $R(\theta, \delta) = E_\theta(||\delta(X) - \xi(\theta)||^2)$). Show that the estimator $\delta(x) = x$ is admissible. Apply this result when (X_1, X_2) are independent normal, independent Poisson, independent binomial, or the sample means from Von-Mises variables. [Using the bivariate information inequality leads to replacement of 4.5(7) by

$$(1) \qquad 2\nabla \cdot b(\theta) + ||b(\theta)||^2 \le 0$$

where $\nabla \cdot b(\theta) = \sum_{i=1}^{2} \frac{\partial b_i(\theta)}{\partial \theta_i}$. If b satisfies (1) so does

(2) $$\bar{b}(\theta) = (2\pi)^{-1} \int_0^{2\pi} Q_\phi^{-1}\, b(Q_\phi \theta)\, d\phi$$

where

$$Q_\phi = \begin{pmatrix} \cos\phi & -\sin\phi \\ \sin\phi & \cos\phi \end{pmatrix} .$$

$\bar{b}$ is spherically symmetric; hence can be written as $\bar{b}(\theta) = \beta(||\theta||)\theta$. Let $t = ||\theta||$. (1) becomes

(3) $$2k\beta(t) + 2t\beta'(t) + t^2\beta^2(t) \leq 0 \quad .$$

Now let $K(t) = t^2\beta(t)$ to get

(4) $$2K'(t) + K^2(t)/t \leq 0$$

in place of 4.5(8). (Note how the argument fails if $k > 2$!)] (Stein (1956), Brown and Hwang (1982, Corollary 4.1).)

4.5.8

Let $X \sim \Gamma(\alpha, \sigma)$, $\alpha > 0$ a specified constant. Consider the problem of estimating $\sigma = -\frac{1}{\theta}$ under the loss function

(1) $$L(\sigma, a) = \frac{a}{\sigma} - \ln(\frac{a}{\sigma}) - 1 \quad .$$

(See Chapter 5 for a natural interpretation of this loss. See also Exercises 4.11.3 and 4.11.4.) Let $\delta_0(x) = \frac{x}{\alpha}$ and let $\delta(x) = (1 + \phi(x))\delta_0(x)$ be any estimator. Let

(2) $$e(\theta) = E_\theta(\phi) \quad \text{and} \quad W(t) = t - \ln(1 + t) , \quad t > -1 \quad .$$

(i) Show that

(3) $$R(\theta, \delta) - R(\theta, \delta_0) \geq -\frac{\theta e'(\theta)}{\alpha} + W(e(\theta))$$

(ii) Use (3) to show that δ_0 is admissible among all estimators having $e(\theta) \leq B$ for all $\theta \in (-\infty, 0)$. (δ_0 is actually admissible with no restriction on δ. See Brown (1966).)

[(i) $R(\theta, \delta) - R(\theta, \delta_0) = E_\theta(-\theta X\phi(X) - \ln(1 + \phi(X)))$
$= -\theta e'(\theta)/\alpha + E_\theta(\phi(X) - \ln(1 + \phi(X)))$. For (ii) follow the pattern of the proof of Theorem 4.5. (It is also possible to use (3) to prove δ_0 is admissible with no restriction on δ.)]

4.6.1

Prove 4.6(2). [Use the information inequality to write

$$\int h(\rho)R(\rho, \delta)\, d\rho \geq m + \int\{2h(\rho)\, \mathrm{Tr}(\nabla b(\rho)) + h(\rho)\mathrm{Tr}(J(\rho)b(\rho)b'(\rho))\}d\rho .$$

Integrate by parts the first term in the integrand in order to get an integral whose integrand is a quadratic in $b(\rho)$ for each fixed ρ. Minimize this integrand for each ρ to get 4.5(2).] See Exercise 5.8.1 for a statistical application of 4.5(2).

4.6.2

In preparation for the proof of 4.6(6) prove the following facts: (i) For each K, $\nabla e_{(K)}(\rho)$ exists for all but at most a countable number of points, ρ.

Fix ρ_0, K for which $\nabla e_{(K)}(\rho_0)$ exists. Let $\delta^*(x) = \delta_{(K)}(x; \rho_0)$, $e^*(\rho) = E_{\theta(\rho)}(\delta^*(X))$, and $D = (d_{ij}) = \nabla e^*(\rho_0) - \nabla e_{(K)}(\rho_0)$. Show

(ii) $d_{ij} = 0$, $i \neq j$, and

(iii) $|d_{ii}| \leq P_{\theta(\rho_0)}(|X_i - \theta_i| \geq K)$.

Let $|D| = (|d_{ij}|)$ with d_{ij} as above and let $J = J(\rho_0)$ be symmetric positive definite with eigenvalues $\lambda_1 \geq \ldots \geq \lambda_m > 0$. Show

(iv) $$\mathrm{Tr}(JDJ^{-1}D) \leq \frac{\lambda_1}{\lambda_m}\mathrm{Tr}|D| \leq \frac{\lambda_1 R_K(\rho_0,\delta)}{\lambda_m K^2} .$$

[Since $|d_{ii}| \leq 1$ the eigenvalues -- and hence diagonal elements -- of JDJ^{-1} all have magnitude at most $\frac{\lambda_1}{\lambda_m}$. Then $R_K = \mathrm{Tr}(J^{-1}E) \geq \frac{1}{\lambda_1}\mathrm{Tr}\, E$ where $E = E_{\rho_0}((\delta - \rho_0)(\delta - \rho_0)')$.]

4.6.3

Also in preparation for the proof of 4.6(6) prove the following matrix inequalities

(i) $Tr(JA'J^{-1}A) \geq 0$ for any (k×k) positive definite symmetric J and any (k× k) matrix A.

(ii) $Tr(J(A' + B')J^{-1}(A + B)) \geq \alpha\, Tr(JA'J^{-1}A) - \frac{\alpha}{1-\alpha} Tr(JB'J^{-1}B)$.

[(i) Diagonalize J (and J^{-1}) and then write out Tr(•) as a sum of individual terms. (ii) follows from (i).]

4.6.4

Now prove 4.6(6). [Write the information inequality for δ^*. Substitute $\nabla e^*(\rho_0) = \nabla e_{(K)}(\rho_0) + D$ and use 4.6.2(iv) and 4.6.3(ii). (Note that both these inequalities are nearly trivial when k = 1, so in that case the overall proof is much simpler to follow.)]

4.6.5

The inequality 4.6(6) is never sharp (except sometimes in the limit as $K \to \infty$). To examine how far from sharp the inequality is compare R_K and the best lower bound from 4.6(6) in the case where k = 1, L is ordinary squared error loss, X N(θ, 1), ρ = θ, and δ(x) = ax (0 < a ≤ 1). [For a = 1, K = 1, I get R_K = .516 ≥ .250 = best lower bound. For a = 1, K = 3 I get R_K = .991 ≥ .5625 and for a = 1, K = 10 R_K = .999+ ≥ .891 .]

4.6.6

Prove 4.6(7). [See 4.6.1.]

4.6.7

Investigate the sharpness of (7) by comparing the Bayes risk for L_K and the bound on the right of 4.6(7) when k = 1, L is ordinary squared error loss, $X \sim N(\theta, 1)$, ρ = θ, and h is a normal $(0, \sigma^2)$ density. (Note: h does not have compact support, but it can be shown (Exercise !) that the tails of h decrease fast enough so that 4.6(7) is still valid.) [When K = ∞

so that 4.6(7) reduces to 4.6(2) the Bayes risk is $\sigma^2/(1 + \sigma^2)$ and the lower bound is $(\sigma^2 - 1)/\sigma^2$. Thus even when $K = \infty$ the bound is not sharp, although it is asymptotically sharp as $\sigma^2 \to \infty$ also.]

4.11.1

Let δ denote the James-Stein estimator 4.11(1) with $r \equiv k - 2$ and let δ^+ denote the corresponding "positive part" estimator 4.11(6). Show that $R(\theta, \delta^+) < R(\theta, \delta)$. [Write $R(\theta, \delta) - R(\theta, \delta^+) = E_\theta(g(||X||^2))$. Note $S^-(g) = 1$ and $IS^-(g) = -1$, and (trivially) $E_0^+(g(||x||^2)) > 0$. Use Exercise 2.21.1.]

4.11.2

Suppose $X \sim N(\mu, \sigma^2 I)$ $(X \in R^k)$ and, independently, $V/\sigma^2 \sim \chi^2_m$. It is desired to estimate μ with squared error loss -- σ^2 is unknown. Let $k \geq 3$. Let $\hat{\sigma}^2 = V/m$ and

$$\delta(x) = \left(1 - \frac{s(||x||^2, \hat{\sigma}^2)}{||x||^2/\hat{\sigma}^2}\right) x$$

where $0 \leq s(\cdot) \leq 2(k-2)m/(m + 2)$ and $s(\cdot, \hat{\sigma}^2)$ is differentiable and non-decreasing for each value of $\hat{\sigma}^2$. Show that $\delta(x)$ is better than $\delta_0(x) = x$. [Assume (w.l.o.g.) that $\sigma^2 = 1$. Condition on $\hat{\sigma}^2$; apply 4.11(5) with $r(\cdot) = \hat{\sigma}^2 s(\cdot, \hat{\sigma}^2)$; and take the expectation over $\hat{\sigma}^2$. (A frequently recommended choice for s is $s(||x||^2, \hat{\sigma}^2) = \min(||x||^2/\hat{\sigma}^2, (k-2)m/(m + 2))$ corresponding to 4.11(6).]

4.11.3

Let $_i$ be independent $\Gamma(\alpha_i, \sigma_i)$ variables with α_i known, $i=1,\ldots,k$. Consider the problem of estimating $\sigma = (\sigma_1,\ldots,\sigma_k)$ with loss function $L(\sigma, a) = \Sigma\sigma_i(1 - a_i/\sigma_i)^2$. The best linear estimator for this problem is δ_0 with $\delta_{0i}(x) = x_i/(\alpha_i + 1)$. (i) When $k = 1$ this estimator is admissible. [Use Theorem 4.5.] (ii) for $k \geq 2$ define δ by

(1) $$\delta_i(x) = x_i/(\alpha_i + 1) + (k-1)\alpha_i + 1/\sum_{j=1}^{k}(\alpha_j + 1)^3/x_j \quad .$$

Show that $R(\sigma, \delta) < R(\theta, \delta_0)$. (This is the easiest of several interesting related results in Berger (1980b).) [Let $\phi_i(x) = (\alpha_i + 1)(\delta_i(x) - \delta_{0i}(x))/x_i$. Using Corollary 4.7 show

(2) $$R(\sigma, \delta_0) - R(\sigma, \delta) =$$

$$-E[\Sigma \left(\frac{2X_i^2}{(\alpha_i + 1)^2} \frac{\partial}{\partial x_i} \phi_i(X) \frac{X_i \phi_i^2}{\alpha_i + 1} + \frac{2X_i^2 \phi_i(X)}{(\alpha_i + 1)^2} \frac{\partial}{\partial x_i} \phi_i(X) \right)]$$

since $\sigma_i(1 - a/\sigma_i)^2 = (a\sqrt{-\theta_i} - 1/\sqrt{-\theta_i})^2$. Then show the expectand on the right of (2) is negative. (Use the fact that $\frac{\partial}{\partial x_i} \phi_i(x) < 0$ to eliminate the terms involving $\phi_i \frac{\partial}{\partial x_i} \phi_i$.)]

4.11.4

Let $X_i \sim \Gamma(\alpha_i, \sigma_i)$, $\alpha_i > 0$ specified constants, $i=1,\ldots,k$, as in Exercise 4.11.3. Consider the loss function

(1) $$L(\sigma, a) = \sum_{i=1}^{k} (a_i/\sigma_i - \ln(a_i/\sigma_i) - 1) .$$

Define δ_0 by $\delta_{0i}(x) = x/\alpha_i$. (See Exercise 4.5.7.) Let $k \geq 3$ and define δ by $\delta_i(x) = (1 + \phi_i(x))\, \delta_{0i}(x)$ where

(2) $$\phi_i(x) = -\frac{c\alpha_i \ln x_i}{1 + \Sigma(\alpha_i \ln x_i)^2}$$

with $0 < c \leq 1$. Show that $R(\sigma, \delta) < R(\sigma, \delta_0)$, $\sigma > 0$. [The unbiased estimator of $R(\sigma, \delta) - R(\sigma, \delta_0)$ is

(3) $$\Sigma[(\frac{x_i}{\alpha_i}) \frac{\partial \phi_i}{\partial x_i} + \phi_i - \ln(1 + \phi_i)] .$$

(The following algebra can be simplified by changing variables in (3) to $y_i = \alpha \ln x_i$, $i=1,\ldots,k$.) Then show this is always positive, using the facts that $|\phi_i| \leq c/2$ and $t - \ln(1 + t) \leq 2t^2/3$ for $|t| \leq ½$. (You will see that values of c somewhat larger than 1 can also be used in (2).)] See Dey, Ghosh, and Srinivasan (1983).

(Change variables in Exercise 4.5.7(3) to $\sigma = -\frac{1}{\theta} = \xi(\theta)$, and compare with the i -th term in brackets in (3), above. This identity of expressions is analogous to that which occurs in the estimation of normal means with squared error loss. See 4.11(8).)

4.11.5

Let $X \sim N(\theta, I)$. Consider the problem of estimating $\theta \in R^k$ under squared error loss. Suppose for some $C < \infty$, $\varepsilon > 0$

(1) $$(\delta_1(x) - x) \cdot x \geq 2 - k + \varepsilon \quad \text{for} \quad ||x|| > C .$$

Then $\delta_1(x)$ is inadmissible.

[Let $\delta_2(x) = \delta_1(x) - \varepsilon[(||x|| - C)^+ \wedge 1] \dfrac{x}{||x||^2}$

and use 4.10(4).] (Note that this generalizes Example 4.11 since $\delta_1(x) = x$ satisfies (1) when $k \geq 3$.)

4.15.1

(i) Show that for estimating the natural parameter the correspondence between prior measures and their generalized Bayes procedures is one-one if Supp ν has a non-empty interior (i.e. show $\delta_G = \delta_H$ a.e.(ν) implies $G = H$). [Use Theorem 4.15 and Corollary 2.13.] (ii) Give an example to show that this unicity may fail if $(\text{Supp } \nu)^\circ = \phi$.

4.15.2

Show that every admissible estimator of θ under squared error loss satisfies the monotonicity condition

(1) $$(x_2 - x_1) \cdot (\delta(x_2) - \delta(x_1)) \geq 0 \quad \text{a.e.}(\nu \times \nu) .$$

[Use 4.14, 4.15, and 2.5. (Do not use 4.16(1) for this would not yield (1) for $x_i \in \partial K$.)]

4.16.1

Let $X \sim P(\lambda)$. Let $c_0 \le 0$. Show that the estimator $\delta(0) = c_0$, $\delta(x) = \ln x$, $x=1,2,\ldots$, is not an admissible estimator of the natural parameter $\theta = \ln \lambda$ under squared error loss. (δ is the "maximum likelihood estimate" of θ; see Chapter 5. Also, the squared error loss function $L(\theta, a) = (a - \theta)^2$ can be justified in its own right, or one can transform to $\lambda = e^{\theta}$ and let $b = e^{a}$. The loss then takes the form $(\ln b - \ln \lambda)^2$ $= (\ln (b/\lambda))^2 = L^*(\lambda, b)$. The inadmissibility result, above, then says also that $\delta^*(x) = x$ is an inadmissible estimator of λ under loss L^*. Losses of the form L^* appear naturally in scale invariant problems; see Brown (1968).)

[Use Theorem 4.16. If δ is of the form 4.16(1) then, by monotonicity,

$$(1) \qquad \ln [x] \le \frac{\lambda_H'(x)}{\lambda_H(x)} \le \ln ([x] + 1), \qquad x \ge 1 .$$

Hence $\lambda_H(x) \to \infty$ as $x \to \infty$ but $\lambda_H(x) = o(e^{\varepsilon x})$ as $x \to \infty$, $\forall\ \varepsilon > 0$. This is impossible by Lemma 3.5 and Exercise 3.5.1.]

4.17.1

Let $X \sim \mathrm{Bin}(n, p)$, $n \ge 3$, and consider the problem of estimating the natural parameter $\theta = \ln (p/(1 - p))$ under squared error loss. Show that the procedure

$$\delta(x) = \begin{cases} -1 & x = 0 \\ 0 & 1 \le x \le n-1 \\ 1 & x = n \end{cases}$$

is admissible. But, δ is not generalized Bayes. (Note that Corollary 4.17 is not valid here because 4.17(1) is not satisfied. Of course, Theorem 4.16 is satisfied with H giving unit mass to the point $\theta = 0$.)

[Let δ' be another estimator. Suppose $\delta'(0) = -1 + \alpha$, $\alpha > 0$. Then

$$\lim_{\theta \to -\infty} |\theta|^{-1}(R(\theta, \delta') - R(\theta, \delta)) = \alpha > 0 .$$

Hence $R(\theta, \delta') \le R(\theta, \delta)$, $\forall\ \theta$, implies (i) $\delta'(0) \le -1$. Similarly (ii) $\delta'(n) \ge 1$. Among all procedures satisfying (i), (ii) δ uniquely minimizes $R(0, \delta)$. Hence δ is admissible. If δ were generalized Bayes the prior G would have to have support {0} by 2.5; but this would imply $\delta(0) = 0 = \delta(n)$.]

4.17.2

Let $Z \sim \Gamma(\alpha,\sigma)$ as in 4.17.3, below. (i) Show that the estimator $\delta_0(x) \equiv 0$ cannot be represented as a generalized Bayes estimator of $\theta = 1/\sigma$. (ii) For $\alpha < 2$ show δ_0 is admissible. (iii) For $\alpha > 2$ show δ_0 is inadmissible. [(ii) If $\delta \ne \delta_0$ then, for some $\epsilon > 0$, $R(\theta,\delta) > \epsilon\theta^\alpha$ as $\theta \to 0$. (iii) Let $\delta(x) = (\alpha-2)/x$.]

4.17.3

Let $Z \sim \Gamma(\alpha, \sigma)$, α known. Then the distributions of $X = -Z$ form an exponential family with natural parameter $\theta = 1/\sigma$. Consider the problem of estimating θ with squared error loss. Show that

(1) $$\delta(x) = be^{x} \quad (= be^{-z})$$

can be represented in the form 4.16(1). [Let H be a Poisson distribution! It can further be shown that δ is admissible when $\alpha \le 2$ since it uniquely minimizes

(2) $$H(\{0\}) \limsup_{\theta\to 0} R(\theta, \delta)e^{\psi(\theta)} + \sum_{i=1}^{\infty} H(\{i\})R(i, \delta)e^{\psi(i)} \quad .]$$

4.17.4

Let $\{p_\theta\}$ be any exponential family with K compact (Binomial, Multinomial, Fisher, Von Mises, etc.). Show that $\delta(x) = x$ is an admissible estimator of θ under squared error loss. [Show that δ is Bayes for the prior distribution, G, with density $c \exp(\psi(\theta) - ||\theta||^2/2)$ and that $B(G) < \infty$. Admissibility then follows from basic decision theoretic results. See, e.g. Lehmann (1983, Theorem 3.1). Use Exercise 3.4.1 to verify that B(G) is finite (and also that G is finite).] Caution! $\delta(x) = x$ is not a very natural estimator of θ, in spite of its admissibility. Hence its use in this problem is not necessarily recommended (unless the prior G is indeed as above). If Supp ν is finite then $\delta(x) = x$ is a natural estimator of $\xi(\theta)$, and is admissible under squared error loss for estimating $\xi(\theta)$. See Exercise 4.5.5 and also Brown (1981b).

4.17.5

Let $X \sim P(\lambda)$. Consider the problem of estimating λ under loss function

(1) $$L(\lambda, a) = (\ln(a/\lambda))^2 .$$

Show that estimator $\delta_1(x) = e^x$ is generalized Bayes, but not admissible. [The question is equivalent to asking whether the estimator $\delta(x) = x$ is generalized Bayes, or admissible for estimating the canonical parameter, θ, under squared error loss. Reason as in Exercise 4.17.4 to show $\delta(x) = x$ is generalized Bayes. However, for estimating θ, direct calculation shows that $\delta'(x) = bx$, $e^{-1} \le b < 1$ is better than $\delta(x)$. This inadmissibility result shows that the general result of Exercise 4.17.4 does not extend to problems with K not compact, even when $k = 1$. (All estimators of the form $\delta(x) = bx$, $0 < b \le 1$, are generalized Bayes for estimating θ. We conjecture that none of them are admissible.)]

4.17.6

Let $X \sim N(\theta, I)$. Consider the problem of estimating $\theta \in R^k$ under squared error loss. (i) Let G be a generalized prior density. Show that the generalized Bayes estimator (if it exists) can be written in the form

(1) $$\delta_G(x) = x + \frac{\nabla g^*(x)}{g^*(x)}$$

where

(2) $$g^*(x) = \int p_\theta(x)\, G(d\theta) .$$

(ii) Consider the linear partial differential inequality

(3) $$\nabla \cdot (g^*(x)\, \nabla u(x)) \leq 0 \qquad ||x|| > 1$$

subject to the condition that u is continuous on $||x|| \geq 1$, and

(4) $$u(x) = 1 \quad \text{for} \quad ||x|| = 1, \qquad u(x) \leq 1 \quad \text{for} \quad ||x|| > 1 \ .$$

Show that if (3), (4) have a non-constant solution which also satisfies

(5) $$E_\theta(||\frac{\nabla u}{u}||^2) < \infty, \qquad \theta \in R^k ,$$

then δ_G is inadmissible.

[(ii) Let $\delta(x) = \delta_G(x) + \frac{\nabla u}{u}$. Use (5) and Green's theorem to justify an expression like 4.10(4) for $R(\theta, \delta_G) - R(\theta, \delta)$ but with an extra term involving a surface integral over $\{x: ||x|| = 1\}$. This extra term is non-negative because of (4), and the remainder of the expression is non-negative because of (3). (Note that Exercise 4.11.5 is a special case of the above. Brown (1971) proves that solubility of (3),(4) implies inadmissibility of δ_G (condition (5) is not required), and conversely if $\frac{\nabla g^*(x)}{g^*(x)}$ is bounded -- and somewhat more generally -- then insolubility of (3), (4) implies admissibility of δ_G. See also Srinivasan (1981).]

<u>4.17.7</u> (Berger and Srinivasan (1978).)

(i) Again let $X \sim N(\theta, I)$ and consider the problem of estimating $\theta \in R^k$ under squared error loss. Suppose

(1) $$\delta(x) = x + \frac{Bx}{x'Mx} + O(\frac{1}{||x||^2})$$

for two constant k×k matrices B and M. Show that δ is inadmissible unless $B = cM$ for some $c \in R$.

[Theorem 4.17 and the representation 4.17.5(1) imply

$$\nabla(\ln g^*(x)) = \frac{Bx}{x'Mx} + O(\frac{1}{||x||^2}) \ .$$

By considering line integrals over closed paths show this is impossible unless B = cM. The calculations are easier if B and M are simultaneously diagonalized (w.l.o.g.). Then when k = 2 the only paths that need be considered are those bounding sets of the form $\{x: x_1 \geq 0, x_2 \geq 0, r \leq ||x|| \leq r + \varepsilon\}$.]

(ii) Suppose, instead, that $X \sim N(\theta, \Sigma)$ with Σ known (positive definite); and δ is given by (1). Now write a necessary condition on B and M for admissiblity of δ. Does the condition involve Σ? What if the loss function is $L(\theta, a) = (a - \theta)' D(a - \theta)$ for some (known positive definite matrix D?

<u>4.18.1</u>

Verify the assertion in 4.18(2). [Use Lemma 3.5 and Exercise 3.5.1(2).]

<u>4.20.1</u>

If $x \notin S$ as defined in 4.20(1) then $\int p_\theta(x)g(\theta)d\theta = \infty$, so that the generalized Bayes procedure for the conjugate prior does not exist at x. [See Exercise 4.19.1.]

<u>4.20.2</u>

Show that Karlin's condition 4.5(2) implies that $S \supset K°$. (Hence, if $\nu(\partial K) = 0$ it implies that the estimator 4.20(2) = 4.5(1) is generalized Bayes.) (ii) Give an example where 4.5(2) is satisfied but 4.20(2) = 4.5(1) is not generalized Bayes.

<u>4.20.3</u>

Let $\{p_\theta: \theta \in \Theta\}$ be a stratum of an exponential family, as defined in Exercise 3.12.1. Suppose it is desired to estimate

$\eta(\theta) = \dfrac{\xi_{(1)}(\theta)}{\xi_{(2)}(\theta)}$ under squared error loss. (Note that in the sequential setting of 3.12.2(iii) and 3.12.3, $\eta(\theta) = E_\theta(Y)$ is a very natural quantity to estimate.) State general conditions to justify the formal manipulation --

$$(1) \qquad \frac{x_{(1)}}{x_{(2)}} = \frac{\int \eta(\theta) e^{\theta \cdot x} \, d\theta_{(1)}}{\int e^{\theta \cdot x} \, d\theta_{(1)}} , \qquad x_{(1)} \in (K_{(1)})^\circ \quad --$$

which says that $\delta(x) = \frac{x_{(1)}}{x_{(2)}}$ is generalized Bayes on $(K_{(1)})^\circ$ relative to the prior measure $d\theta_{(1)}$ on $\Theta^* = \{(\theta_{(1)}, \theta_{(2)}(\theta_{(1)})) \in \Theta : \theta_{(1)} \in (N_{(1)})^\circ\}$. (The conclusion is justified in the situation of 3.12.2(ii) and in that of 3.12.2(iii) if $p_\theta(N \le N_0) = 1$, and somewhat more generally.)

4.20.4

Generalize 4.20.3(1) to obtain a representation for certain estimators of the form $\frac{x_{(1)} + a}{x_{(2)} + b}$.

4.21.1

Show that 4.21(5) and 4.21(2) imply 4.21(1) and 4.21(3). [4.21(1) is trivial from 4.21(5'). For 4.21(3) reason as in the proof of Theorem 4.19. The key fact is that, with q as defined there,

$$\int_{-B}^{q} -\xi_1(\theta_1,\theta_2,\ldots,\theta_k) g(\theta) e^{\theta\cdot x - \psi(\theta)} \, d\theta_1 = - \int_{-B}^{q} \left(x_1 g(\theta) + \frac{\partial g(\theta)}{\partial \theta_1}\right) e^{\theta\cdot x-\psi(\theta)}$$

$$+ \; g(\theta) e^{\theta\cdot x - \psi(\theta)} \Big]_{-B}^{q} , \quad \text{etc.}$$

Now integrate over $\theta_2,\ldots,\theta_k$ and let $B \to \infty$. The first part of the expression is bounded because of 4.21(1), (2), and the second part because of 4.21(5').]

4.21.2 (Converse to Theorem 4.19.)

Let G be a prior measure whose Bayes procedure for estimating $\xi(\theta)$ exists on S and satisfies $\delta(x) = \alpha x + \beta$. Suppose $S^\circ \neq \phi$. Assume further that G possesses a density g satisfying 4.21(1), (2), (3). Then G is a conjugate prior measure, and its conjugate prior density, 4.18(1), has $\alpha = 1/(\lambda + 1)$ and $\beta = \gamma/(\lambda + 1)$.) Apply 4.7(3) to the last integral of the equality

$$(1)\qquad (\lambda+1)\int(\nabla\psi(\theta))g(\theta)e^{\theta\cdot x-\psi(\theta)}d\theta \;=\; \gamma\int g(\theta)e^{\theta\cdot x-\psi(\theta)}d\theta + x\int g(\theta)e^{\theta\cdot x-\psi(\theta)}d\theta \;,$$

rearrange terms and invoke completeness to find

$$(2)\qquad \nabla g(\theta) \;=\; (\gamma - \lambda\nabla\psi(\theta))g(\theta) \qquad .]$$

(Diaconis and Ylvisaker (1979) show that this statement is true without this "further" assumption that G possess a density.)

(A question of interest is whether this unicity result extends to non-linear generalized Bayes estimators. To be more precise suppose the generalized Bayes procedures for estimating $\xi(\theta)$ under priors G and H exist and are equal everywhere on S with $S^\circ \neq \phi$. Does this imply G = H? In the case of the normal distributions or the Poisson distribution the answer is yes. See 4.15.1 for the normal distribution and Johnstone (1982) for the Poisson distribution.)

<u>4.24.1</u>

Suppose $\delta(\cdot)$ is admissible for estimating ξ under squared error loss. Then $\nu\{x : \delta(x) \notin K\} = 0$.

[Define $\delta'(x)$ as the projection of $\delta(x)$ on K. If $\nu\{x : \delta(x) \neq \delta'(x)\} \neq 0$ then $R(\theta, \delta') < R(\theta, \delta)$ whenever $R(\theta, \delta) < \infty$. (If δ is admissible there must exist some θ for which $R(\theta, \delta) < \infty$.)]

<u>4.24.2</u>

(i) Verify that the conclusion of Theorem 4.24 remains valid when $\{p_\theta\}$ is a steep exponential family and $\Theta \subset N^\circ$. (ii) Even more generally, it is valid for any one-parameter exponential family if

$$(1)\qquad \Theta \;\subset\; \{\theta : E_\theta(X) = \xi(\theta) \in R\}$$

and if the definition 4.24(1) is modified to

(2) $$I'_\delta = \{x : \nu(\{y: y > x,\ \delta(y) \in \xi(N^\circ)^\circ\}) > 0$$

and $$\xi(\{y: y < x,\ \delta(y) \in \xi(N^\circ)^\circ\}) > 0\} .$$

(iii) Extend Theorem 4.24 to the problem of estimating $\rho(\theta)$ under squared error loss where $\rho: N^\circ \to R$ is a non-decreasing function. [The formulation and proof are identical to (ii), above.]

4.24.3

Let $\nu = \nu_1 + \nu_2$ where ν_1 is Lebesgue measure on (0, 3) and ν_2 gives mass 1 to each of the points x = 1,2. Consider the estimator δ of ξ (under squared error loss) given by

(1) $$\delta(x) = \begin{cases} 0 & x < 1 \\ \tfrac{1}{2} & x = 1 \\ 1\tfrac{1}{2} & 1 < x < 2 \\ 2\tfrac{1}{2} & x = 2 \\ 3 & x > 2 \end{cases} .$$

(i) Show that δ has the representation 4.24(2) on $I = (1,2)$, but (ii) this representation cannot be extended to the points x = 1,2 even though $\delta(x) \in K^\circ$ for these points. (iii) Show that δ is a pointwise limit of a sequence of Bayes procedures. (δ is also admissible. See Exercise 7.9.1.)

4.24.4

Let X have the geometric distribution with parameter p(Ge(p)), under which

(1) $$\Pr\{X = x\} = p(1 - p)^x \qquad x=0,1,\ldots \quad .$$

(i) Show that $\delta(x) = x/2$ is an admissible estimator of $E_p(X) = (1-p)/p$ under squared error loss. [Use Karlin's Theorem 4.5. Note also that the estimators $\delta(x) = cx$ with $c > \tfrac{1}{2}$ fail to satisfy 4.5(2) and are not admissible.] (ii) Suppose it is known in addition that $p \le \tfrac{1}{2}$, so that $E_p(X) \ge 1$. Using Theorem 4.24 show that the truncated version of δ-- namely $\delta'(x) = \max(\delta(x), 1)$ --

is inadmissible. (iii) Can you find an (admissible) estimator better than δ' ??)

CHAPTER 5. MAXIMUM LIKELIHOOD ESTIMATION

5.1 Definition

Let $\phi : R^k \to [0, \infty]$ be convex. Define $\ell : R^k \times R^k \to [-\infty, \infty]$ by

(1) $$\ell(\theta, x) = \ell_\phi(\theta, x) = \theta \cdot x - \phi(\theta) \quad .$$

For $S \subset N$ let

(2) $$\ell(S, x) = \sup \{\ell(\theta, x) : \theta \in S\}$$

and let

(3) $$\hat{\theta}_S(x) = \{\theta \in S : \ell(\theta, x) = \ell(S, x)\} \quad .$$

Note that according to this definition $\hat{\theta}_S$ is a subset of S. We will often abuse the notation slightly by letting $\hat{\theta}$ also denote an element of this set.

If $\phi = \psi$ is the cumulant generating function for an exponential family then

$$\ell_\psi(\theta, x) = \log p_\theta(x) \qquad \theta \in N$$

is the *log likelihood function* on N. (Of course, $\ell_\psi(\theta, x) \equiv -\infty$ for $\theta \notin N$ in accordance with the natural convention that $\psi(\theta) = \infty$ for $\theta \notin N$.) $\hat{\theta} \in \hat{\theta}_S(x)$ is then called a maximum *likelihood estimate* at x relative to $S \subset N$. A function $\delta : K \to \Theta$ for which $\delta(x) \in \hat{\theta}_\Theta(x)$ a.e.(ν) is called the (a) *maximum likelihood estimator*. This terminology is not always properly used in the literature; and we will also abuse it, at least to the extent of also referring to the set valued function $\hat{\theta}(\cdot)$ as the maximum likelihood estimator.

5.2 Assumptions

The main results of this section concern the existence and construction of maximum likelihood estimators, $\hat{\theta}$. The proofs of these results are based on the fact that ψ is a convex function satisfying certain additional properties, and not otherwise on the fact that ψ is a cumulant generating function. In Chapter 6 we will want to apply these same existence and construction results to convex functions, ϕ, which are not cumulant generating functions. To prepare for this application we now make explicit the conditions on ϕ which are needed in the proofs of the main results of this section.

Let $\phi : R^k \to (-\infty, \infty]$ be a lower semicontinuous convex function. Let $N = N_\phi = \{\theta : \phi(\theta) < \infty\}$. Such a function is called *regularly strictly convex* if it is strictly convex and twice differentiable on N_ϕ°, and

(1) $$D_2\phi \quad \text{is positive definite on} \quad N_\phi^\circ \quad .$$

In the following results we will assume ϕ is regularly strictly convex. In some of the following we also assume ϕ is steep. Note that if ψ is the cumulant generating function of a steep exponential family then it satisfies these assumptions.

Here are some useful facts.

Let $\ell = \ell_\phi$ be defined by 5.1(1), and let the mapping $\xi : N_\phi \to R^k$, be defined by $\xi(\theta) = \nabla\phi(\theta)$. Then, ξ is continuous and 1 - 1 since ϕ is strictly convex. (1) says that the Hessian of $\xi = \nabla\phi$ is positive definite. Hence $\xi(N^\circ)$ is an open set; call it R, or R_ϕ. $\xi^{-1}(\cdot)$ is continuous on R.

Theorem 3.6 establishes that

(2) $$R = K^\circ$$

when $\phi = \psi$ is the cumulant generating function of a minimal steep exponential family. In particular, in this case

(3) $$R \quad \text{is convex} \quad .$$

It will be shown in Proposition 6.7 that (3) is always valid under the above general assumptions on ϕ including steepness of ϕ.

As previously, let $\theta(\cdot) = \xi^{-1}(\cdot)$, i.e. $\xi(\theta(x)) = x$.

(The assumption above of the existence of second derivatives and of (1) is convenient, but can be dispensed with. The other assumptions are required for the following development.)

We emphasize again: the following results about ℓ_ϕ and maximum likelihood estimation concern the general situation where ϕ is as assumed above. These results therefore apply in particular to *maximum likelihood estimation from minimal steep standard exponential families.*

5.3 Lemma

Assume ϕ is regularly strictly convex. Then, $\ell(\cdot, x)$ is concave and upper semicontinuous on R^k for all $x \in R^k$. It is strictly concave on N.

If $\theta_0 \in N^\circ$ then

(1) $$\nabla \ell(\cdot, x)_{|\theta_0} = x - \xi(\theta_0)$$

(2) $$D_2\ell(\cdot, x)_{|\theta_0} = -D_2\phi(\theta_0) = -\not{Z}(\theta_0)$$

where $(\not{Z}(\theta_0))_{ij} \quad \dfrac{\partial^2}{\partial\theta_i\partial\theta_j}\phi(\theta_0)$ is positive definite. If $x \in R$ $(= K^\circ)$ then

(3) $$\lim_{||\theta||\to\infty} \ell(\theta, x) = -\infty \quad .$$

Proof. The first assertions are immediate from Assumption 5.2. Equations (1) and (2) are a direct calculation. The positive definiteness of $\not{Z}(\theta_0)$ is a consequence of 5.2(1).

Assertion (3) has been proved in 3.6(4) for the case where $\phi = \psi$ is the cumulant generating function of a minimal steep exponential family. This proof was needed in order to show that $R = K^\circ$ in such a situation. However we now want a proof valid for arbitrary convex functions, ϕ, satisfying

5.2(1). This is easily supplied.

Assume $x \in R$, then $\theta(x) \in N^\circ$. Note using (1), (2) that $\nabla \ell(\theta(x), x) = 0$, and $D_2 \ell(\theta(x), x)$ is negative definite. Hence for some $\delta > 0$, $\varepsilon > 0$

$$(4) \qquad \ell(\theta, x) = \ell(\theta(x), x) - (\theta - \theta(x))' \not\Sigma(\theta - \theta(x))/2 + o(||\theta - \theta(x)||^2)$$

$$< \ell(\theta(x), x) - \varepsilon \qquad \text{for} \qquad ||\theta - \theta(x)|| = \delta \quad .$$

It follows that when $||\theta - \theta(x)|| > \delta$

$$(5) \qquad \ell(\theta, x) \leq \ell(\theta(x), x) - \frac{||\theta - \theta_0||}{\delta}\varepsilon$$

by (4) since

$$\ell(\theta(x) + (\delta/(||\theta - \theta(x)||))(\theta - \theta(x))) \leq (1 - \delta/||\theta - \theta(x)||)\ell(\theta(x), x)$$
$$+ (\delta/||\theta - \theta(x)||)\ell(\theta, x)$$

by convexity. (5) implies (3). ||

(We note that the positive definiteness of $\not\Sigma$ is not really needed to establish (3). It is only necessary that the conclusion of (4) be valid -- i.e. for some $\delta > 0$, $\varepsilon > 0$

$$(4') \qquad \ell(\theta, x) < \ell(\theta(x), x) - \varepsilon \qquad \text{for} \qquad ||\theta - \theta(x)|| = \delta \quad .$$

This condition follows whenever $\ell(\cdot, x)$ is a strictly concave function which assumes its maximum at $\theta(x)$.)

It is useful to now prove the following lemma. This result is used in Theorem 5.5 to show that $\hat{\theta}_\Theta \subset N^\circ$ when Θ is convex.

5.4 Lemma

Assume ϕ is steep and regularly strictly convex. Let $\theta_1 \in N - N^\circ$, $\theta_0 \in N^\circ$. Let $\theta_\rho = \theta_0 + \rho(\theta_1 - \theta_0)$, $0 < \rho < 1$. Then

(1) $$\lim_{\rho \uparrow 1} \left(\frac{\partial}{\partial \rho} \ell(\theta_\rho, x) \right) = -\infty \quad .$$

Hence there is a $\rho' < 1$ such that

(2) $$\ell(\theta_{\rho'}, x) > \ell(\theta_1, x) \quad .$$

Proof. From 5.3(1)

$$\frac{\partial}{\partial \rho} \ell(\theta_\rho, x) = (\theta_0 - \theta_1) \cdot (x - \xi(\theta_\rho)) \to -\infty$$

as $\rho \uparrow 1$ because ψ is steep. This proves (1) from which (2) is immediate. (In case ψ is regular, i.e. $N = N^\circ$, then $\lim_{\rho \uparrow 1} \ell(\theta_\rho, x) = -\infty$ by upper semi-continuity, which can also be used to prove (2).) ||

FULL FAMILIES

Here is a fundamental result concerning maximum likelihood estimation. It follows easily from the above.

5.5 Theorem

Let ϕ be steep and regularly strictly convex. If $x \in R$ then

(1) $$\hat{\theta}_N(x) = \{\theta(x)\} \subset N^\circ \quad .$$

In other words, $\hat{\theta}_N(x)$ consists of the unique point $\hat{\theta} = \theta(x)$ satisfying

(1') $$\xi(\hat{\theta}) = x \in R \quad .$$

If $x \notin R$ then $\hat{\theta}_N(x)$ is empty. (Recall that if $\phi = \psi$ is the cumulant generating function of a steep canonical exponential family then $R = K^\circ$.)

Proof. For any x, $\{\hat{\theta}_N(x)\} \subset N^\circ$ by virtue of Lemma 5.4. Any maximum likelihood estimator must thus be a local maxima of $\ell(\cdot, x)$ and hence must satisfy

$$\nabla \ell(\cdot, x)_{|\hat{\theta}} = 0 \quad .$$

This implies (1') by 5.3(1). Furthermore, the solution to (1') is unique if it exists, and it exists if and only if $x \in R = \xi(N^\circ)$. ||

Remarks. Maximum likelihood estimation is defined in statistical theory for a general parametric family of densities $\{f_\theta : \theta \in \Theta\}$ by $\hat{\theta}(x) = \{\theta \in \Theta : f_\theta(x) = \sup_\alpha f_\alpha(x)\}$. Note that this definition is invariant under reparametrization. Thus, if $\xi = \xi(\theta)$ is a 1 - 1 map on Θ the maximum likelihood estimate of the parameter $\xi \in \xi(\Theta)$ is $\xi(\hat{\theta})$.

Accordingly, Theorem 5.5 says that for minimal steep exponential families $x = \xi(\theta(x))$ is the unique maximum likelihood estimator of the mean value parameter $\xi = \xi(\theta)$ at $x \in K^\circ$. To emphasize, in terms of the mean value parametrization the maximum likelihood estimator is determined by the trivial equation

$$(1'') \qquad \hat{\xi}(x) = x , \qquad x \in K^\circ \quad .$$

For the present, (1") is valid if and only if $x \in K^\circ$. This set of course contains almost every $x(\nu)$ if and only if

$$(2) \qquad \nu(K - K^\circ) = 0 \quad .$$

Note that (2) is satisfied if ν is absolutely continuous with respect to Lebesgue measure. It is never satisfied if ν has finite support or, more generally, has countable support and $K \neq R^k$. In the last part of Chapter 6 we expand such exponential families so that (1") usually remains valid for a.e.x (ν).

(Since $\xi = E_\theta(x)$ equation (1") also defines $\hat{\xi}(x) = x$ as the classical method-of-moments estimator. Thus for the mean value parametrization the maximum likelihood and method-of-moments estimators agree.)

Suppose that $X_1, \ldots, X_n$ are independent identically distributed random variables from the exponential family $\{p_\theta\}$. Then, as noted in 1.11(2), the distributions of the sufficient statistic $\bar{X}_n = n^{-1} \sum_{i=1}^{n} X_i$ also form an exponential family with natural parameter $\alpha = n\theta$ and cumulant generating function $n\psi(\alpha/n)$. It follows that $\alpha(x) = n\theta(x)$. So, the maximum likelihood estimator of α based on $\bar{X}_n$ is $n\theta(\bar{X}_n)$ and the maximum likelihood estimator $\hat{\theta}_{(n)}$ of $\theta = \alpha/n$ based on $\bar{X}_n$ is

(3) $$\hat{\theta}_{(n)} = \hat{\alpha}/n = \theta(\bar{x}_n) \ .$$

5.6 Examples (Beta Distribution)

For a variety of common full families the above remarks lead to easy calculation of the maximum likelihood estimator. These are situations such as those mentioned in 3.8 where the mean value parametrization has a convenient form. For example if $Y_1, Y_2, \ldots, Y_n$ are i.i.d. multivariate normal (μ, Σ) random variables then the maximum likelihood estimators for μ and $\mu\mu' + \Sigma$ are, respectively, $\bar{Y} = n^{-1} \sum_{i=1}^{n} Y_i$ and $n^{-1} \sum_{i=1}^{n} Y_i Y_i'$. This leads to the conventional maximum likelihood estimates

(1) $$\hat{\mu} = \bar{Y}$$
$$\hat{\Sigma} = S = n^{-1} \Sigma(Y_i - \bar{Y})(Y_i - \bar{Y})' \ .$$

For the Fisher - Von Mises distributions the result of Theorem 5.5 is not so easy to implement. See 3.8. Another not so convenient, but important, family is the beta family, which will now be discussed.

Consider the family of densities

(2) $$f_{\alpha,\beta}(y) = B^{-1}(\alpha, \beta) y^{\alpha-1}(1 - y)^{\beta-1}, \quad 0 < x < 1, \quad \alpha > 0, \quad \beta > 0 \ .$$

realtive to Lebesgue measure on (0, 1), where $B = B(\alpha, \beta)$ denotes the beta function,

(3) $$B(\alpha, \beta) = \frac{\Gamma(\alpha)\Gamma(\beta)}{\Gamma(\alpha + \beta)}$$

This is a two parameter exponential family with canonical parameters $(\alpha, \beta) \in N = (0, \infty) \times (0, \infty)$. The corresponding canonical statistics are

$$(4) \qquad x_1 = \log y \qquad x_2 = \log (1 - y) \quad .$$

In this case the canonical parameters themselves have a convenient statistical interpretation since

$$(5) \qquad E(Y) = \alpha/(\alpha + \beta), \qquad E(1 - Y) = \beta/(\alpha + \beta)$$

$$\operatorname{Var}(Y) = \alpha\beta/(\alpha + \beta)^2(\alpha + \beta + 1) = \operatorname{Var}(1 - Y) \quad .$$

The mean value parameters are somewhat less convenient. One has

$$(6) \qquad \xi_2(\beta, \alpha) = \xi_1(\alpha, \beta) = B^{-1}(\alpha, \beta) \int_0^1 (\ln y) y^{\alpha-1}(1 - y)^{\beta-1} dy$$

$$= \frac{\partial}{\partial\alpha} (\ln B(\alpha, \beta)) = \frac{\Gamma'(\alpha)}{\Gamma(\alpha)} - \frac{\Gamma'(\alpha + \beta)}{\Gamma(\alpha + \beta)}$$

$$= - \sum_{k=0}^{\infty} \left(\frac{1}{\alpha+k} - \frac{1}{\alpha+\beta+k}\right) = - \sum_{k=0}^{\infty} \frac{\beta}{(\alpha+k)(\alpha+\beta+k)} \quad ,$$

and

$$(7) \qquad \xi_1(\alpha, \beta) = - \sum_{k=0}^{\beta-1} \frac{1}{\alpha+k} \qquad \text{if} \qquad \beta = 1,2,\ldots \quad .$$

(See e.g. Courant and Hilbert (1953, p.499)).

Suppose $Y_1,\ldots,Y_n$ are i.i.d. beta variables, and X_{1i}, X_{2i} are defined from Y_i through (3), $i=1,\ldots,n$. Then the maximum likelihood estimates of (α, β) can be found numerically by solving

$$(8) \qquad \xi_j(\hat{\alpha}, \hat{\beta}) = \bar{X}_j \qquad j = 1, 2$$

from (6), where $\bar{X}_j = n^{-1} \sum_{i=1}^{n} X_{ji}$. An exact solution appears to be unavailable, except when $\hat{\alpha},\hat{\beta}$ turn out to be integers so that (7) applies.

According to Theorem 5.5, the solution to (8) exists if and only if $\bar{X} \in K^\circ$. Now,

$$K = \text{conhull}\{\ln y, \ \ln(1-y) : y \in (0, \infty)\} \quad .$$

Since $\{\ln y, \ln(1-y) : y \in (0, 1)\}$ is strictly convex in R^2 this solution therefore exists if and only if $n \geq 2$ and $\sum_{i=1}^{n} (Y_i - \bar{Y})^2 > 0$. The event $\sum_{i=1}^{n} (Y_i - \bar{Y})^2 = 0$ occurs with zero probability when $n \geq 2$; hence the maximum likelihood estimate exists with probability one when $n \geq 2$.

NON-FULL FAMILIES

We now proceed to discuss the existence and construction of maximum likelihood estimators when $\Theta \subsetneq N$. Here is an existence theorem.

5.7 Theorem

Let ϕ be steep and regularly strictly convex. Let $\Theta \subset N$ be a non-empty relatively closed subset of N. Suppose $x \in R$. Then $\hat{\theta}_\Theta(x)$ is non-empty.

Suppose $x \in \bar{R} - R$. Suppose there are values $x_i \in R$, $i=1,\ldots,I$, and constants $\beta_i < \infty$ such that

$$\Theta \subset \bigcup_{i=1}^{I} H^-((x - x_i), \beta_i) \quad . \tag{1}$$

Then $\hat{\theta}_\Theta(x)$ is non-empty.

Remark. See Exercises 5.7.1-2, 7.9.1-3, and Theorem 5.8 for more information about the theorem. In particular, (1) implies $x \notin (\xi(\Theta))^-$. See Figure 5.7(1) for an illustration of 5.7(1).

Proof. Let $x \in R$. $\ell(\cdot, x)$ is upper semi-continuous and satisfies 5.3(3). Hence $\ell(\cdot, x)$ assumes its supremum over $\bar{\Theta}$. But $\ell(\theta, x) = -\infty$ for $\theta \in (\bar{\Theta} - \Theta) \subset \bar{N} - N$. It follows that $\hat{\theta}_\Theta(x)$ is non-empty.

Suppose $x \in \bar{R} - R$ and (1) is valid. Then for each $\theta \in \Theta$ there is an index i for which $\theta \in \bar{H}^-(x - x_i, \beta_i)$. For this index

$$\ell(\theta, x) = \theta \cdot (x - x_i) + \theta \cdot x_i - \psi(\theta) \leq \beta_i + \theta \cdot x_i - \psi(\theta) \quad . \tag{2}$$

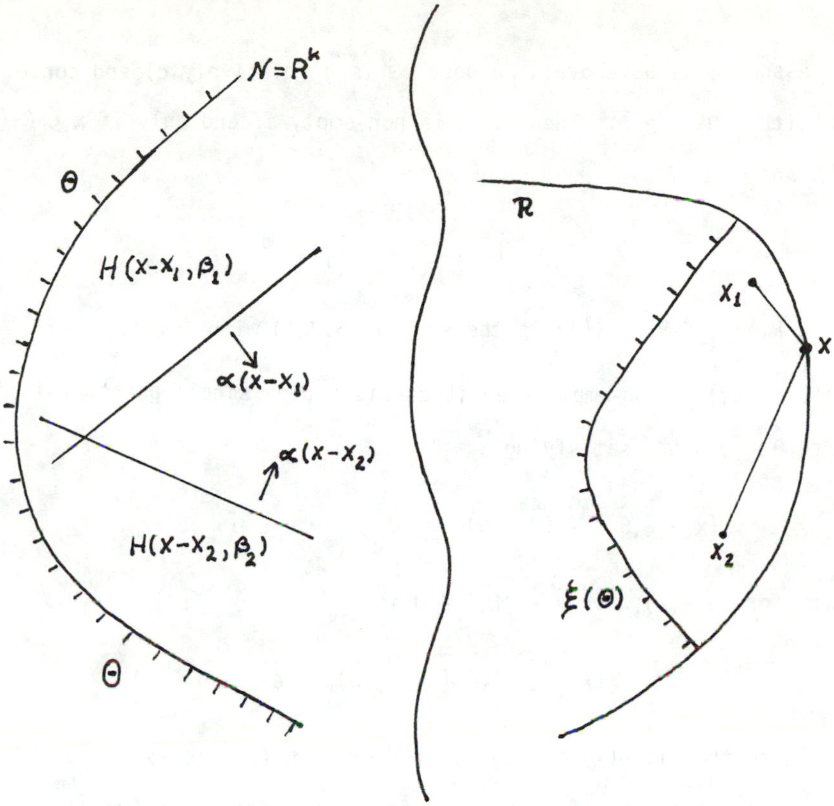

Figure 5.7(1):

An Illustration of 5.7(1) showing $\mathcal{R}$, $x \in \bar{\mathcal{R}} - \mathcal{R}$, $\Theta \subset \bigcup_{i=1}^{2} H^-((x - x_i), \beta_i)$ and $\xi(\Theta)$.

It follows that

$$\ell(\theta, x) \le \sup \{\beta_i + \theta \cdot x_i - \psi(\theta) : 1 \le i \le I\} \to -\infty \text{ as } ||\theta|| \to \infty ,$$

$$\theta \in \Theta ,$$

by 5.3(3). The second assertion of the theorem follows from (2) as did the first from 5.3(3). ||

CONVEX PARAMETER SPACE

When Θ is convex one gets a better result, including a fundamental equation defining the maximum likelihood estimator.

5.8 Theorem

Assume ϕ is as above. Suppose Θ is a relatively closed convex subset of N with $\Theta \cap N^\circ \neq \phi$. Then $\hat{\theta}_\Theta(x)$ is non-empty if and only if $x \in R$ $(= K^\circ)$ or $x \in \bar{R} - R$ and

(1) $$\Theta \subset H^-(x - x_1, \beta_1)$$

for some $x_1 \in R$, $\beta_1 \in R$. ((1) is the same as 5.7(1) with $I = 1$.)

If $\hat{\theta}_\Theta(x)$ is non-empty then it consists of a single point. This is the unique point, $\hat{\theta} \in \Theta \cap N^\circ$ satisfying

(2) $$(x - \xi(\hat{\theta})) \cdot (\hat{\theta} - \theta) \geq 0 \qquad \forall\, \theta \in \Theta \quad .$$

(An alternate form of (2) when $x - \xi(\hat{\theta}) \neq 0$ is

(2') $$\Theta \subset \bar{H}^-(x - \xi(\hat{\theta}),\ (x - \xi(\hat{\theta})) \cdot \hat{\theta}) \quad .)$$

(Note that if $\theta(x) \in \Theta$ then, of course, $\hat{\theta}_\Theta(x) = \{\theta(x)\}$ and $\hat{\theta} = \theta(x)$ trivially satisfies (2). See 5.9 for illustrations of (2).)

Proof. $\ell(\cdot, x)$ is strictly concave on N and hence can assume its maximum at only one point of the convex set Θ. Furthermore, $\hat{\theta}_\Theta \subset N^\circ$ by Lemma 5.4.

Suppose (2) is satisfied. Then for $\theta \in \Theta$

(3) $$\begin{aligned} \ell(\hat{\theta}, x) - \ell(\theta, x) &= (\hat{\theta} - \theta) \cdot (x - \xi(\hat{\theta})) \\ &\quad + (\hat{\theta} - \theta) \cdot \xi(\hat{\theta}) - (\psi(\hat{\theta}) - \psi(\theta)) \\ &= (\hat{\theta} - \theta) \cdot (x - \xi(\hat{\theta})) + \ell(\hat{\theta}, \xi(\hat{\theta})) - \ell(\theta, \xi(\hat{\theta})) \\ &\geq 0 + 0 = 0 \ , \end{aligned}$$

with equality if and only if $\theta = \hat{\theta}$. $(\ell(\hat{\theta}, \xi(\hat{\theta})) - \ell(\theta, \xi(\hat{\theta})) > 0$ when $\theta \neq \hat{\theta}$ since $\hat{\theta} = \theta(\xi(\hat{\theta}))$ is the unique maximum likelihood estimator over N corresponding to the observation $\xi(\hat{\theta})$.) Hence (2) implies that $\hat{\theta}_\Theta(x) = \{\hat{\theta}\}$.

On the other hand, suppose

(4) $$(x - \xi(\theta_0)) \cdot (\theta_0 - \theta_1) < 0 \qquad \text{for some } \theta_0, \ \theta_1 \in \Theta \quad .$$

Then

$$\theta_\rho = \theta_0 + \rho(\theta_1 - \theta_0) \in \Theta$$

for $0 \le \rho \le 1$ since Θ is convex. Then

$$\frac{d}{d\rho} \ell(\theta_\rho, x)_{|\rho=0} = (x - \xi(\theta_0)) \cdot (\theta_1 - \theta_0) > 0 \quad .$$

Hence $\ell(\theta_\rho, x) > \ell(\theta_0, x)$ for $\rho > 0$ sufficiently small; and θ_0 cannot be the unique maximum likelihood estimator. It follows that the unique maximum likelihood estimator if it exists, must satisfy (2).

Finally, if $x \in R$ or (1) is satisfied then $\hat{\theta}_\Theta$ is non-empty by Theorem 5.7. Conversely, if $\hat{\theta} \in \hat{\theta}_\Theta$ is non-empty then $\hat{\theta}$ satisfies (2). Hence $\hat{\xi} = \xi(\hat{\theta}) \in R$ and

$$(x - \hat{\xi}) \cdot \theta \le (x - \hat{\xi}) \cdot \hat{\theta}$$

by (2) so that (1) is satisfied with $x_1 = \hat{\xi}$. ||

5.9 Construction

The criterion 5.8(1) is particularly easy to apply if $\Theta = (\theta_0 + L) \cap N$ for some linear subspace, L. This is because the vectors $\{(\hat{\theta} - \theta) : \theta \in L\}$ will then span L. Thus, by (1), in order to find $\hat{\theta}$ one need only search for the unique point $\theta^* \in \Theta$ for which $x - \xi(\theta^*) \perp L$. This process can be viewed from two slightly different perspectives. Because of its importance we illustrate both these perspectives in the simplest case where $\theta_0 + L$ is a hyperplane.

Thus, consider the case where $\Theta = H \cap N$ with H a hyperplane, say $H = H(a, \alpha)$. Let $x \in R$. (The same construction also works for $x \in \bar{R} - R$ if 5.7(1) is satisfied.) To find $\hat{\theta}_\Theta(x)$ one may proceed from $\theta(x)$ along the curve $\{\theta(x + \rho a) : \rho \in R\}$ until the unique point at which $\theta(x + \rho a) \in \Theta$. This point is $\hat{\theta}$. The process is illustrated in Figure 5.9(1).

An alternative procedure is to map $\Theta \cap N^\circ$ into R as $\xi(\Theta \cap N^\circ)$. Then proceed along the line $\{x + \rho a : \rho \in R\}$ until the unique point at which $x + \rho a \in \xi(\Theta \cap N^\circ)$. This point is $\hat{x} = \xi(\hat{\theta})$. This process is illustrated in Figure 5.9(2).

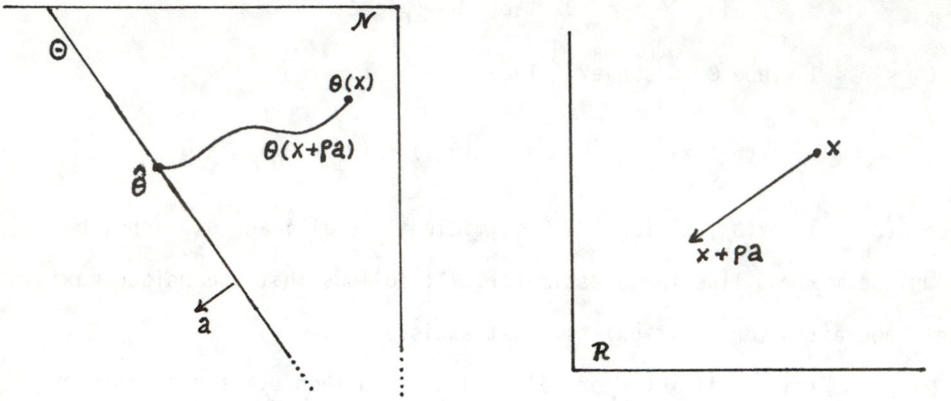

Figure 5.9(1): First construction of $\hat{\theta}$

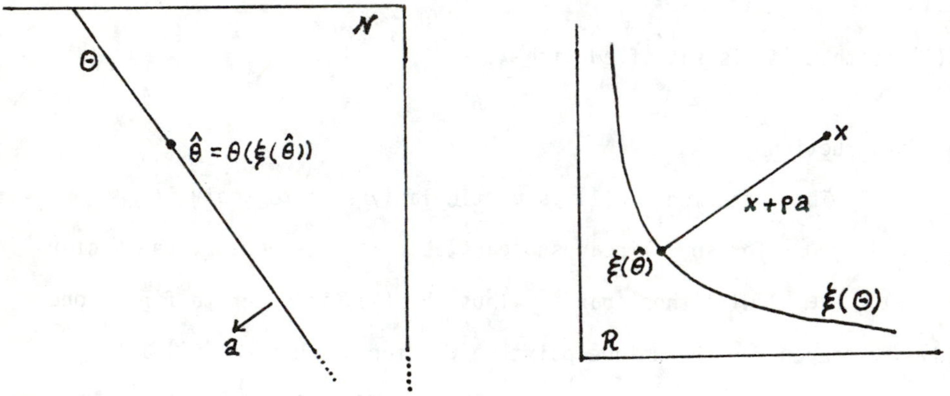

Figure 5.9(2): Second construction of $\hat{\theta}$

There are useful paradigms available also for the case where Θ is an arbitrary relatively closed convex set. These are described in 5.13.

The entire process illustrated above may also be viewed from a different perspective. Θ is contained in a proper linear subset of R^k. Hence the densities $\{p_\theta : \theta \in \Theta\}$ form an exponential family which is not minimal. This non-minimal family can be reduced by sufficiency and reparametrization to a minimal family of dimension $k' < k$. Let $(\phi_1, \ldots, \phi_{k'})$ and

$(y_1,\ldots,y_{k'})$ denote the natural parameters and corresponding observations in this family. (They are formed by projecting θ and x, respectively, onto $H(a, \alpha)$ or any translate $H(a, \beta)$.) This family will have log-Laplace transform $\psi^*(\phi) = \psi(\theta(\phi))$, and the m.l.e., $\hat{\phi}$, satisfies 5.5(1) -- i.e.

$$\hat{\phi}(y) = \phi(y)$$

where $\phi(y)$ is the inverse to $\xi^*(\phi) = \nabla\psi^*(\phi)$. Thus

$$\hat{\theta}(x) = \phi(y(x)) \quad .$$

These remarks can be used to yield a very simple proof of Theorem 5.8 in the special case where $\Theta = (\theta_0 + L) \cap N$. They also provide a method of easily constructing the maximum likelihood estimate in many such cases. Here are two examples.

5.10a Example

Consider the classical Hardy-Weinberg situation described in Example 1.8. (X_1, X_2, X_3) is multinomial (N, ξ) with expectation $\xi = N(p^2, 2pq, q^2)$, $0 < p = 1-q < 1$. This is a three-dimensional exponential family with two dimensional parameter space

$\Theta = \{\theta: = \beta_1(1,1,1) + \beta_2(2,1,0) + (0, \ln 2, 0)\} = H((1,-2,1), -2 \ln 2)$.

(This family is not minimal. This fact affects but does not hinder the reasoning which follows.)

Reduction to a minimal exponential family yields a one-parameter exponential family with parameter $\phi = 2\theta_1 + \theta_2$ and natural observation $y = 2x_1 + x_2$. (Θ is two-dimensional but yields a family of only order one since the original family was not minimal.) Note that

$$E(Y) = N(2p^2 + 2pq) = 2pN \quad . \tag{1}$$

Hence

$$\hat{p} = \frac{y}{2N} = \frac{2x_1 + x_2}{2N} , \qquad 0 < y < 2N \quad . \tag{2}$$

Correspondingly, $\hat{\xi} = N(\hat{p}^2, 2\hat{p}\hat{q}, \hat{q}^2)$ and $\hat{\theta}$ can be defined from $\hat{\theta}_i = \beta_1 + \ln \hat{\xi}_i$, $\beta_1 \in R$. (Note that $\hat{\theta}$ is a line rather than a single point because the original representation of the multinomial family was not minimal.)

The simplicity of (1) is the special fact which enables the preceding construction to proceed so smoothly. Many other multinomial log-linear models behave similarly. Classes of such models are discussed in Darroch, Lauritzen, and Speed (1980) and in Haberman (1974). Here is a useful example.

5.10b Example

Consider a 2×2×2 contingency table. The observations will be denoted by y_{ijk}, $i,j,k = 0,1$. They are multinomial (N) variables with respective probabilities π_{ijk}. There are various useful log-linear models for such a table. The derivation of maximum likelihood estimates for such models provides a useful and illuminating application of the preceding theory. Here we consider the model in which responses in the first category (corresponding to index i) are conditionally independent of those in the third category given the level of response in the second category. This model illustrates several characteristic phenomena, and allows for direct and explicit maximum likelihood estimates of the parameters π_{ijk}.

In order to write the model in customary vector-matrix notation, let $z_\ell = y_{ijk}$ where $\ell = 1 + i + 2j + 4k$ $(1 \le \ell \le 8)$, and, similarly, $\pi_\ell = \pi_{ijk}$. Let $(\log \pi)$ denote the vector with coordinates $\log \pi_\ell$, $\ell=1,\dots,8$. Let

$$D' = \begin{matrix} 1 & -1 & 1 & -1 & 1 & -1 & 1 & -1 \\ 1 & 1 & -1 & -1 & 1 & 1 & -1 & -1 \\ 1 & 1 & 1 & 1 & -1 & -1 & -1 & -1 \\ 1 & -1 & -1 & 1 & 1 & -1 & -1 & 1 \\ 1 & 1 & -1 & -1 & -1 & -1 & 1 & 1 \\ 1 & 1 & 1 & 1 & 1 & 1 & 1 & 1 \end{matrix}$$

The log-linear model of interest here has

$$\theta^* = (\log \pi) = D\beta, \qquad \beta \in R^6 \quad . \tag{2}$$

In order to normalize π one must choose β_6 so that

(3) $$\sum_{\ell=1}^{8} \pi_\ell = 1 \quad .$$

The resulting multinomial family is an 8-parameter exponential family. Its canonical statistic can be reduced via sufficiency. Let $x^* = D'z$ so that $\theta^{*\prime}z = \beta'D'z = \beta'x^*$. Furthermore $x_6^* = N$ with probability one. Hence $x \in R^5$ with $(x_1,\ldots,x_5) = (x_1^*,\ldots,x_5^*)$ is a sufficient, canonical statistic. The corresponding canonical parameter is $\theta \in R^5$ with $(\theta_1,\ldots,\theta_5) = (\beta_1,\ldots,\beta_5)$. It can be checked thatthis log-linear family is characterized by the conditional independence of responses in categories 1 and 3 given level of response in category 2. The conditional independence can be checked by noting that if $i \neq i'$, $k \neq k'$ then (2) yields

$$\ln \pi_{i'jk'} + \ln \pi_{ijk} = \ln \pi_{i'jk} + \ln \pi_{ijk'} \quad .$$

From this it follows that $\pi_{+j+}\,\pi_{ijk} = \pi_{+jk}\,\pi_{ij+}$, which implies the desired conditional independence.

By Theorem 4.5 x is the maximum likelihood estimate of $E(X) = \xi(\theta)$. Thus, $(\log \pi) = D\beta(x)$ is the maximum likelihood estimate of $(\log \pi) = \theta^*$ with $\beta(x) = (\beta_1(x),\ldots,\beta_5(x), \beta_6(\beta))'$ where $\beta_6(\cdot)$ is determined by (3).

The relation between $\xi(\theta)$ and $\pi(\theta)$ is easy to determine via simple calculations such as $\xi_1 = E(X_1) = \Sigma(-1)^i\, E(y_{ijk})$, etc. These yield

(4) $$\begin{aligned} \xi_1 &= N\Sigma(-1)^i\pi_{ijk} & \xi_2 &= N\Sigma(-1)^j\pi_{ijk} \\ \xi_3 &= N\Sigma(-1)^k\pi_{ijk} & \xi_4 &= N\Sigma(-1)^{i+j}\pi_{ijk} \\ \xi_5 &= N\Sigma(-1)^{j+k}\pi_{ijk} \quad . \end{aligned}$$

Thus

(5) $$\Sigma(-1)^i y_{ijk} = x_1 = N\Sigma(-1)^i\hat{\pi}_{ijk} \quad , \qquad \text{etc.}$$

From these relationships and the structure of D it is possible in this case to give explicit expressions for $\{\hat{\pi}_{ijk}\}$ in terms of $\{y_{ijk}\}$. Let a "+" replacing a subscript denote addition over that subscript. Thus, $\pi_{1++} = \sum_{j,k} \pi_{1jk}$. Simple manipulation based on (3) and (5) yields

$$N\hat{\pi}_{i++} = y_{i++}, \qquad N\hat{\pi}_{+j+} = y_{+j+}, \qquad N\hat{\pi}_{++k} = y_{++k}$$

(6)

$$N\hat{\pi}_{ij+} = y_{ij+}, \qquad N\hat{\pi}_{+jk} = y_{+jk} \quad .$$

The conditional independence properties yield

$$\pi_{ijk} = \pi_{+j+} \frac{\pi_{ij+}}{\pi_{+j+}} \frac{\pi_{+jk}}{\pi_{+j+}} = \pi_{ij+}\,\pi_{+jk}/\pi_{+j+} \quad .$$

Hence

(7) $$N\hat{\pi}_{ijk} = y_{ij+}\, y_{+jk}/y_{+j+} \quad .$$

FUNDAMENTAL EQUATION

5.11 Definition

For $\theta_0 \in \Theta \subset R^k$ define $\nabla_\Theta(\theta_0)$, the *set of (outward) normals* to Θ at $\theta_0 \in \Theta$, to be the set of all $\delta \in R^k$ satisfying

(1) $$\delta \cdot (\theta_0 - \theta) \geq 0 + o(||\theta_0 - \theta||) \quad \forall \ \theta \in \Theta \quad .$$

∇ is obviously a convex cone, and can easily be shown to be closed.

Note that if $\theta_0 \in \text{int}\,\Theta$ then $\nabla_\Theta(\theta_0) = \{0\}$. If θ_0 is an isolated point of Θ then $\nabla_\Theta(\theta_0) = R^k$. If Θ is a differentiable manifold with tangent space T at θ_0 then $\nabla_\Theta(\theta_0)$ is the orthogonal complement of T -- i.e., $\nabla_\Theta(\theta_0) = \{\delta: \delta \cdot \tau = 0 \ \ \forall \tau \in T\}$. Here $\nabla_\Theta(\theta_0)$ is a linear subspace of R^k. If Θ is convex and $\theta_0 \in \text{bd}\,\Theta$ then $\nabla_\Theta(\theta) = \{\delta : \Theta \subset \bar{H}^-(\delta, \delta\cdot\theta_0)\}$.

5.12 Theorem

Assume ϕ is steep and regularly strictly convex. Let Θ be a relatively closed subset of N. Then for any $\hat{\theta} \in \hat{\theta}_\Theta(x) \cap N^\circ$

(1) $$x - \xi(\hat{\theta}) \in \nabla_\Theta(\hat{\theta}) \quad .$$

Proof. Let $\hat{\theta} \in \hat{\theta}_\Theta(x) \subset N^\circ$. Note that

(2) $$\nabla_\theta(\ell(\theta, \xi(\hat{\theta}))_{|\theta=\hat{\theta}} = 0 \quad .$$

and $x - \xi(\theta) = 0$ when $\theta = \hat{\theta}$. Hence, as in 5.8(3)

(3) $$0 \leq \ell(\hat{\theta},x) - \ell(\theta,x) = (\hat{\theta} - \theta) \cdot (x - \xi(\hat{\theta})) + \ell(\hat{\theta}, \xi(\hat{\theta})) - \ell(\theta, \xi(\hat{\theta}))$$

$$= (\hat{\theta} - \theta) \cdot (x - \xi(\hat{\theta})) + o(||\hat{\theta} - \theta||) \quad .$$

Thus, by definition, (1) is satisfied. ||

Note that the theorem does not require $x \in R$ (= K°).

5.13 Construction

The fundamental equation, 5.8(1) or 5.12(1), can be used to picture the process of finding a maximum likelihood estimator, by an extension of the process pictured in 5.9.

Fix $x \in R^k$. Suppose it is desired to locate $\hat{\theta}_\Theta(x)$. If $\Theta \cap N^\circ \neq \phi$ one should first check to see whether $x \in \xi(\Theta \cap N^\circ)$. If so, then $\theta(x) = \hat{\theta}_\Theta(x)$. If not, then $\hat{\theta}_\Theta(x) \subset \text{bd}\ \Theta$. To see whether a given $\theta_0 \in \text{bd}\ \Theta \cap N^\circ$ can be an element of $\hat{\theta}$ first locate Θ, θ_0,x, and $x_0 = \xi(\theta_0)$ on their respective graphs. Then carry a vector δ pointing in the direction of $x - x_0$ over to θ_0 in order to check whether δ is an outward normal to Θ at θ_0. If so, then θ_0 is a candidate for $\hat{\theta}$. In fact, if Θ is convex $\{\theta_0\} = \hat{\theta}_\Theta(x)$. If Θ is not convex one must search over bd Θ for all such candidates, then examine $\ell(\theta, x)$ at each of them to eliminate those which are not global maxima. (If ϕ is not regular and Θ is not convex one needs also to search over

$\Theta \cap (N - N^{\circ})$.) The process is illustrated in Figure 5.13(1).

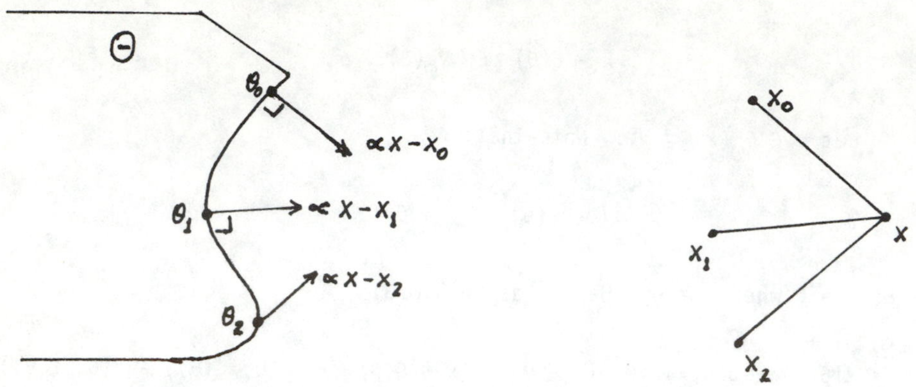

Figure 5.13(1): θ_0 and θ_1 are candidates for $\theta_\Theta(x)$. θ_2 is not.

If bd Θ is a curve as in Figure 5.13(1) then this process is relatively convenient. Otherwise, it is usually less convenient to search over all of bd Θ for the set of candidates.

An alternate picture can also be constructed. In this picture one constructs for each $\theta \in \Theta$ the collection of points in X space for which θ can possibly be the maximum likelihood estimator. In order to construct this picture one locates $\theta \in$ bd Θ and draws the unit outward normal(s), δ, to θ. One then maps θ to $\xi(\theta)$ and carries the vector(s) δ directly over to X space. The corresponding line or cone with vertex located at $\xi(\theta)$ is the locus of values of x for which $\theta \in \hat{\theta}_\Theta(x)$ is a possibility. Again, if x falls in more than one such locus then $\ell(\theta, x)$ must be separately examined at all such θ. This process is illustrated in Figure 5.13(2).

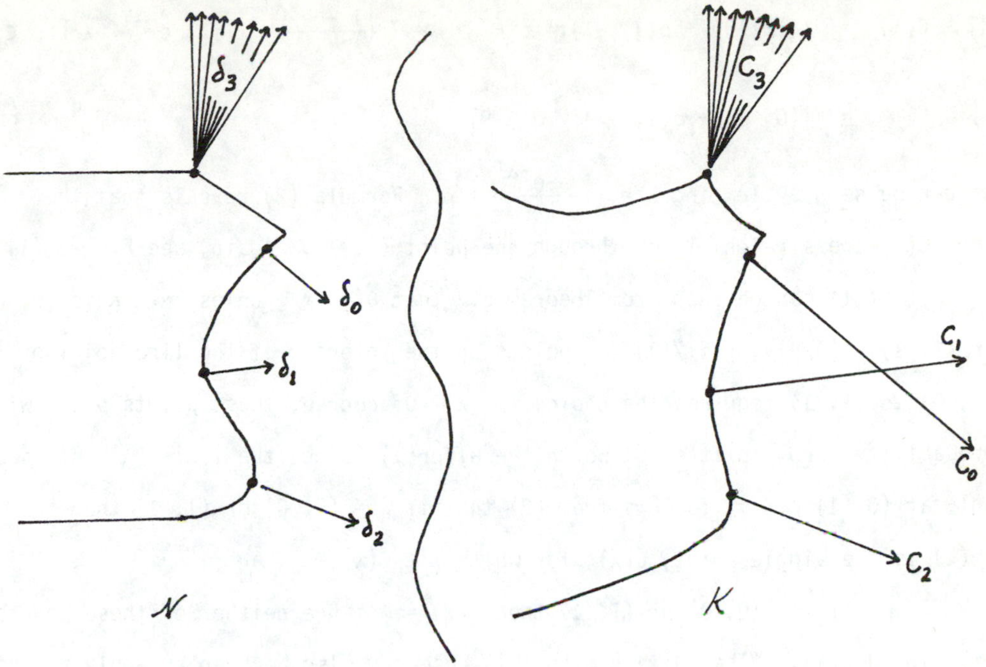

Figure 5.13(2): C_i is the locus of points, x, for which θ_i can possibly fall in $\hat{\theta}_\Theta(x)$.

5.14 Example

The curved exponential family described in Example 3.12 provides a particularly elegant instance of the above construction. The family is a two-parameter standard exponential family with $\theta(\lambda) = (-\lambda, -\ln\lambda)'$, and $\Theta = \{\theta(\lambda): \lambda > 0\} \subset N = (-\infty, 0) \times R$, and $\psi(\theta) = \ln[(e^{\theta_1 T} - 1)/\theta_1 + e^{\theta_1 T + \theta_2}]$. $K = \text{conhull}\ \{(0, 0), (T, 0), (T, 1)\}$.

Then, $\xi(\theta(\lambda)) = (\frac{1 - e^{-\lambda T}}{\lambda}, e^{-\lambda T})$. Figure 5.14(1) shows both Θ and K and $\xi(\Theta)$ on a single plot. There is no overlap since $\Theta \subset \{(\theta_1, \theta_2): \theta_1 < 0\}$ and $K \subset \{(x_1, x_2): x_1 \geq 0\}$.

The tangent space to $\theta(\lambda)$ is spanned by $(-1, -1/\lambda)'$. Hence $\nabla_\Theta(\theta(\lambda))$ is the line $\{\rho(1, -\lambda): \rho \in R\}$. The locus, $C(\lambda)$, of points x for which $\theta(\lambda)$ can be the maximum likelihood estimator is the line

(2) $$C(\lambda) = \{\xi(\theta(\lambda)) + \rho(1, -\lambda): \rho \in R\} = \{(\frac{1 - e^{-\lambda T}}{\lambda} + \rho, \; e^{-\lambda T} - \lambda\rho): \rho \in R\}$$

$$= \{(0, 1) + \sigma(1, -\lambda): \sigma \in R\}$$

as can be seen by letting $\sigma = \frac{1 - e^{-\lambda T}}{\lambda} + \rho$. Formula (2) reveals that the loci $C(\lambda)$ are straight lines through the point (0, 1). Again, see Figure (1).

It can be seen from Theorem 5.7 that $\hat{\theta}(x) \neq \phi$ unless $x \in K$ is (0, 0) or (T, 1). (Applying 5.7(1) for points on the interior of the line joining (0, 0) to (T, 1) requires the choice I = 2. Of course, these points occur with probability zero, so it's not worth the effort!) Since the loci $C(\lambda)$ intersect only at $(0, 1) \notin K$ it follows from (2) that if $x \neq (0, 0)$ or (T, 1) then $\hat{\theta}_\Theta(x)$ is the single point, $\theta(\lambda)$, for which $x \in C(\lambda)$.

If x = (0, 0) or (T, 1) then $\hat{\theta}(x) = \phi$ since neither of these points lies in $\bigcup_{\lambda \in R} C(\lambda)$. (That $\hat{\theta}(x) = \phi$ in this case can also be seen by applying the final part of Theorem 5.8 to the parameter set consisting of the convex hull of Θ).

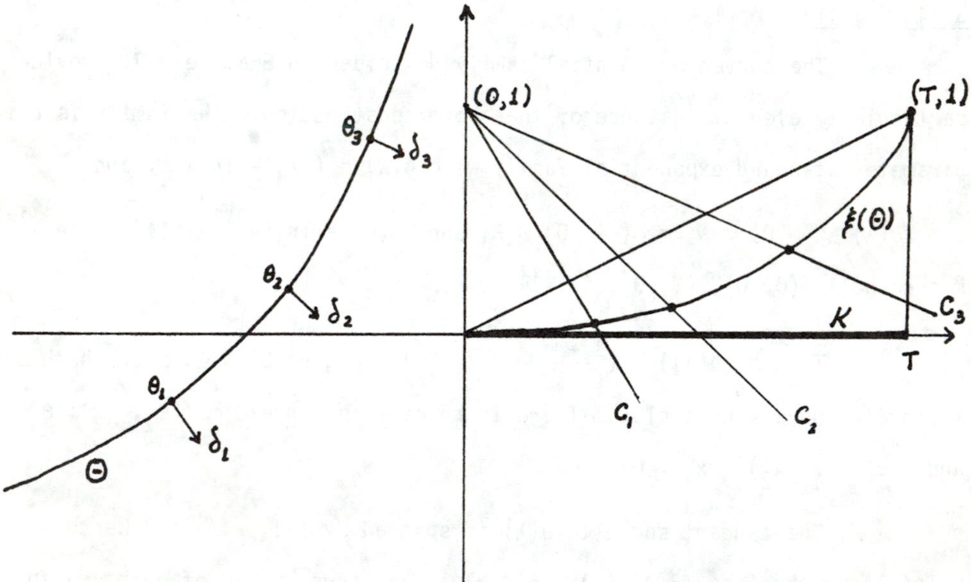

Figure 5.14(1): Illustrating the construction of $\hat{\theta}(x)$ via construction of loci C.

The original description of this example involves a single observation, X, which can take only values in $(0 \times [0, T]) \cup \{(T, 1)\}$. However, if one observes $\bar{X}_n = n^{-1} \sum_{i=1}^{n} X_i$ where X_i are n i.i.d. variables each with the given distribution, then $\bar{X}_n$ can take values over more of K. This problem has natural parameter $\theta^* = n\theta$ and log Laplace transform $\psi^*(\theta^*) = n\psi(\theta^*/n)$. It follows that N, K *and* $\xi(\Theta)$ are as before. Θ undergoes a simple transformation. It is easy to check that the above picture applies equally well to this problem, for which various values of $x \in K°$ are possible. See also Proposition 5.15.

From (2) one sees that the maximum likelihood estimator of λ is

$$\hat{\lambda} = (1 - \bar{x}_2)/\bar{x}_1 \quad . \tag{3}$$

In terms of the original motivation for this problem the parameter $1/\lambda$ is the mean value (= mean lifetime) of the exponential variable Z. Thus,

$$\widehat{(1/\lambda)} = \bar{x}_1/(1 - \bar{x}_2) = \frac{n\bar{x}_1}{n(1 - \bar{x}_2)} \quad . \tag{3'}$$

In this problem $n\bar{x}_1 = \sum_{i=1}^{n} Y_i$ = "total time on test", and $n(1 - \bar{x}_2)$ = (number of observations $< T$) = "number of objects failing before truncation". This supplies the familiar expression for this problem:

$$\widehat{(1/\lambda)} = \frac{\text{total time on test}}{\text{number of objects failing before truncation}} \quad . \tag{3''}$$

Note that the value of T does not appear in (3"). This fact has been commented on and exploited by Cox (1975) and many others.

It has been noted that the differentiable subfamily treated in this example is a stratum within the full two parameter family. It is really this fact which explains the elegance of the above construction and of Figure 5.14(1). See Exercise 5.14.1 - 5.14.3.

In general the maximum likelihood estimate for an i.i.d. sample is determined exactly as that from a single observation. The latter part of Example 5.14 mentions one special case of this. It is worthwhile to formally note this fact.

5.15 Proposition

Let $X_1,\ldots,X_n$ be i.i.d. random variables from a standard exponential family $\{p_\theta: \theta \in \Theta\}$. Let $\hat{\theta}_\Theta$ denote the set of maximum likelihood estimators of $\theta \in \Theta$ on the basis of a single observation.

The maximum likelihood estimator of $\theta \in \Theta$ based on the sample $X_1,\ldots,X_n$ is a function of the sufficient statistic, $\bar{X}_n = n^{-1}\sum_{i=1}^{n} X_i$. Let $\hat{\theta}_\Theta^{(n)}(\cdot)$ denote this function of $\bar{X}_n$. Then

$$\hat{\theta}_\Theta^{(n)}(\bar{x}) = \hat{\theta}_\Theta(\bar{x}) \quad . \tag{1}$$

Proof. The cumulant generating function for the sufficient statistic $S_n = n\bar{X}_n$ is $n\psi(\theta)$. The proposition follows from the fact that

$$\begin{aligned} \ell_{n\psi}(\theta, s) &= \theta \cdot s - n\psi(\theta) \\ &= n(\theta \cdot s/n - \psi(\theta)) \\ &= n\ell_\psi(\theta, s/n) \quad , \end{aligned}$$

since this shows that $\ell_{n\psi}(\cdot, s)$ is maximized if and only if $\ell_\psi(\cdot, s/n)$ is maximized. ||

EXERCISES

5.6.1

Verify formula 5.6(6).

5.6.2

The multivariate generalization of the beta distribution is the *Dirichlet distribution*, $\mathcal{D}(\alpha)$, defined as follows: $k \geq 2$; $\theta_i > 0$, $i=1,\ldots,k$, $\theta_0 = \sum_{i=1}^{k} \theta_i$, $Y_i > 0$, $i=1,\ldots,k$; $Y_k = 1 - \sum_{i=1}^{k-1} Y_i$; the distribution has density with respect to Lebesgue measure over the allowable $\{(y_1,\ldots,y_{k-1})\}$

$$(1) \qquad f_\theta(y) = \frac{\Gamma(\theta_0)}{\prod_{i=1}^{k} \Gamma(\theta_i)} \prod_{i=1}^{k} y_i^{(\theta_i - 1)} .$$

This is a k-parameter exponential family with canonical statistic $X_i = \ln Y_i$.

(i) Describe K.

(ii) Verify the standard formulae:

$$(2) \qquad E(Y_i) = \theta_i/\theta_0 \qquad \operatorname{Var}(Y_i) = \frac{(\theta_0 - \theta_i)\theta_i}{\theta_0^2(\theta_0 + 1)}$$

$$\operatorname{Cov}(Y_i, Y_j) = -\frac{\theta_i\theta_j}{\theta_0^2(\theta_0 + 1)} , \qquad i \neq j .$$

(iii) Derive formulae for $E(X_i)$ analogous to 5.6(6), (7).

(iv) Let $1 = s_0 < \ldots < s_\ell = k$ and define $z_j = \sum_{i=s_{j-1}+1}^{s_j} Y_i$, $j=1,\ldots,\ell$. Show that Z has a $\mathcal{D}(\theta')$ distribution, and describe θ' in terms of θ.

(v) Let $Y^{(i)}$, $i=1,\ldots,n$ be independent, k-dimensional $\mathcal{D}(\theta^{(i)})$ variables. Verify that the distribution of $n^{-1} \sum_{i=1}^{n} Y^{(i)}$ is $\mathcal{D}(\Sigma\theta^{(i)})$.

5.6.3

Let X_i, $i=1,\ldots,k$, be independent $\Gamma(\alpha_i, \beta)$ variables. Describe the conditional distribution of the variable $(X_1,\ldots,X_k)$ given $\sum_{i=1}^{k} X_i$ as a multiple of an appropriate Dirichlet variable. (Note the partial analogy between the situation here and that in Example 1.16. Note also that the situation here was described from another perspective in Exercise 2.15.1.)

5.6.4

The following is a valid statement: the k-dimensional Dirichlet distributions form the family of (proper) conjugate priors for the parameter $(p_1,\ldots,p_{k-1})$ of a k-dimensional multinomial distribution. Relate this statement to the general theory of Sections 4.18-4.20, and describe (in terms of the Dirichlet parameters) the posterior expectation of p given the multinomial observation. [Let $p_i = e^{\theta_i}/(1 + \sum_{j=1}^{k-1} e^{\theta_i})$, etc.]

(This conjugate relation between Dirichlet and multinomial distributions has an infinite dimensional generalization in which the Dirichlet distribution is replaced by a "Dirichlet process" and the multinomial distribution is replaced by a distribution over the family of cumulative distribution functions on [0, 1]. See Ferguson (1973) and Ghosh and Meeden (1984).)

5.7.1

(i) Show that 5.7(1) implies $x \notin (\xi(\Theta))^-$.

(ii) Show the converse is not valid by constructing an example in which $\phi = \psi$, $R = K^\circ$ is not strictly convex, $x \notin (\xi(\Theta))^-$, and 5.7(1) fails. (I believe no example exists when R is strictly convex. See Exercise 7.9.2 which shows that when $R = K^\circ$ is strictly convex and $x \notin (\xi(\Theta))^-$ then $\hat{\theta}(x) \neq \phi$.)

[(i) $x \notin (\xi(H^-((x - x_i), \beta_i)))^-$ for $x_i \in R$, $x \in \bar{R} - R$.

(ii) Let ν give mass 1 to each of the four points $(\pm 1, \pm 1)$. Let $x = (1, 0)$

and $\Theta = \{(t, 2): t \in R\}$.]

5.7.2

Construct examples in which $\phi = \psi$ is steep, $R = K^\circ$, $x \in (\xi(\Theta))^-$, and (i) $\hat{\theta}(x) = \phi$, (ii) $\hat{\theta}(x) \neq \phi$. [For both examples let ν be the uniform distribution on the ball $\{x: (x_1 - 1)^2 + \sum_{i=2}^{k} x_i^2\}$ plus a point mass at 0. For (i) let $\Theta = \{\theta: \theta = (\alpha, 0, \ldots, 0)\}$. For (ii) let $\Theta = \{\theta : \psi(\theta) = 3\}$. For every unit vector $v \neq e$, there is a unique $\eta(v) > 0$ such that $\psi(\eta(v)v) = 3$. As $v \to e_1$, $\eta(v) \to \infty$ and hence $\xi(\eta(v)v) \to 0$. Hence $0 \notin (\xi(\theta))^-$.]

5.8.1

Let $\{p_\theta: \theta \in \Theta\}$ be a standard one-parameter exponential family. Suppose $\xi(\Theta)$ is an unbounded interval -- i.e. $\xi(\Theta) \supset (\xi_0, \xi_1)$ with $\xi_0 = -\infty$ or $\xi_1 = +\infty$. For $\xi_0 < A < \xi_1$ suppose either

$$(1) \qquad \xi_0 = -\infty \quad \text{and} \quad \int_{\xi_0}^{A} J(\xi)d\xi = \infty$$

or

$$\xi_1 = \infty \quad \text{and} \quad \int_{A}^{\xi_1} J(\xi)d\xi = \infty$$

with $J^{-1}(\xi) = \theta'(\xi) = \mathrm{Var}_{\theta(\xi)}(X)$, so that J denotes the Fisher information for estimating ξ. Consider the problem of estimating ξ under the loss 4.6(1) -- i.e. $L(\xi, \delta) = J(\xi)(\delta - \xi)^2$. Show that: (i) the maximum likelihood estimator is minimax; and (ii) if $\Theta \subsetneq N$ then the maximum likelihood estimator is not admissible. (iii) Give examples when $\Theta = N$ and $\xi(\Theta)$ is unbounded in which the maximum likelihood estimator is not minimax, is minimax but not admissible, is both minimax and admissible. (iv) Can you generalize (i) to a k-parameter family?

[Let $\alpha_n \downarrow \xi_0$, $\beta_n \uparrow \xi_1$ and

$$(2) \qquad h_n^{1/2}(\xi) = \min\left(\int_{\alpha_n}^{\xi} J(t)dt,\ K_n,\ \int_{\xi}^{\beta_n} J(t)dt\right)$$

where K_n is chosen so that h_n is a probability density. Show $K_n \to 0$ because of (1). Then use 4.6(2). For (ii) use Theorem 4.24.]

5.9.1

Consider the general linear model as defined in 1.14.1. (a) Verify that the usual least squares estimators of ξ are also the maximum likelihood estimators (i.e. $\hat{\mu} = B\hat{\xi}$). (b) What is the maximum likelihood estimator of σ^2? Is it unbiased? (Assume $m \geq r + 1$.) (c) Generalize the preceding questions to the situation where $Y \sim N(\mu, \sigma^2 \Sigma)$ with $\mu = B\xi$ as in 1.14.1 and Σ a known positive matrix. [The maximum likelihood estimates are the usual generalized least squares estimates.]

5.9.2

Generalize 5.9.1 to the multivariate linear model defined in 1.14.3.

5.9.3

Let (X_1, X_2) be the canonical statistics from a normal sample with mean μ_1 and variance σ_1^2; and let (Z_1, Z_2) be from an independent normal sample with mean μ_2 and variance σ_2^2. Suppose $\mu_1 \leq \mu_2$, but the parameters are otherwise unrestricted. Show that $(\hat{\mu}_1, \hat{\mu}_2) = (x_1, z_1)$ if $x_1 \leq z_1$ and otherwise $\hat{\mu}_1 = \hat{\mu}_2 = \hat{\mu}$ is the unique solution to

$$\frac{x_1}{x_2 - \hat{\mu}^2} + \frac{z_1}{z_2 - \hat{\mu}^2} = \left(\frac{1}{x_2 - \hat{\mu}^2} + \frac{1}{z_2 - \hat{\mu}^2}\right)\hat{\mu} .$$

(Assume $x_2 > x_1^2$ and $z_2 > z_1^2$, which occurs with probability one.)

5.9.4

Let ξ be a normally distributed vector with mean 0 and covariance matrix Σ. Given ξ let Y be distributed according to the general linear model 1.14.1. (Assume $m \geq r + 1$.) Suppose $B'B$ is diagonal and $\Sigma \subset \mathcal{D}$, a relatively closed convex subset of positive definite diagonal matrices.

(a) Show that the (marginal) distributions of Y form an exponential family with Θ a relatively closed proper convex subset of N. [$\hat{\xi}$ and $|(Y - B\hat{\xi})|^2$ are minimal sufficient statistics.] (b) When $\mathcal{D}$ is all positive definite diagonal matrices describe the maximum likelihood estimates of Σ, σ^2. (c) Extend (b) to include other suitable subsets, $\mathcal{D}$. (d) The preceding is a canonical form for a class of random effects models (see, e.g., Arnold (1981)). To see this convert the usual balanced one-way or two-way random effects models to a model of this form by applying suitable linear transformations to the usual parameters. [For the one-way model having $E(Y_{ij}) = \mu + \alpha_i$, $\mu \sim N(0, \sigma_\mu^2)$, $\alpha_i \sim N(0, \sigma_\alpha^2)$, $i=1,\ldots,I$, $j=1,\ldots,J$ let $\xi_1 = I\mu + \sum_{i=1}^{I} \alpha_i$ and $(\xi_2,\ldots,\xi_I) = (\alpha_1,\ldots,\alpha_I)M$ where M is a $I \times (I - 1)$ matrix whose columns are orthonormal and orthogonal to $\underset{\sim}{1}$.]

[The following three exercises concern the $2\times2\times2$ contingency table.]

5.10.1

Consider the model under which the first category and third category are (marginally) independent (i.e., $\pi_{i+k} = \pi_{i++}\pi_{++k}$). Show this is a log-linear model and find an explicit expression for the maximum likelihood estimator.

5.10.2

Consider the log-linear model described by the restriction $0 = \phi_1 + \phi_4 + \phi_6 + \phi_7 - (\phi_2 + \phi_3 + \phi_5 + \phi_8)$. (This is the model described by the phrase, "no third-order interactions.") Write the equation(s) determining the maximum likelihood estimator. Determine that these equations do not have a closed form solution, such as 5.10(7). (See Darroch, Lauritzen, and Speed (1980). In such a case the likelihood equations must be solved numerically. The usual methods are the E-M algorithm or the Newton-Raphson algorithm. See Bishop, Feinberg, and Holland (1975) and Haberman (1974).)

5.12.1

Consider the model described by $\frac{1}{2} = \pi_{0++} = \pi_{1++} = \pi_{+0+} = \pi_{+1+}$.

Show this corresponds to a differentiable subfamily within the full exponential family, but is not a log-linear model. Find the maximum likelihood estimator for this differentiable subfamily [$\pi_{00+} = \pi_{11+}$.]

5.14.1

Let $\{p_\theta: \theta \in \Theta\}$ be a stratum of regular (or a steep) exponential family, as defined in Exercise 3.12.1. (a) Show that for $x \in R$ the maximum likelihood estimator exists and satisfies

$$(1) \qquad \frac{\hat{\xi}_{(1)}}{\hat{\xi}_{(2)}} = \frac{x_{(1)}}{x_{(2)}} .$$

(b) Discuss the situation when $x \in \bar{R} - R$. (c) Show (by example) that there can be two solutions to (1); but there can never be more than two. Is it possible for both of these solutions to be maximum likelihood estimators? [Suppose the family is defined by $\psi(\theta) = \psi_0$. Note that the set $\{\theta: \psi(\theta) \leq \psi_0\}$ is convex and apply Theorem 4.8.]

5.14.2

Show how the result of Exercise 5.14.1 directly yields 5.14(3'). [Translate x_2.]

5.14.3

Apply 5.14.1 to describe the maximum likelihood estimator for the other examples discussed in 3.12.2.

5.15.1

Let $X_1,\ldots,X_n$ be i.i.d. with distribution p_θ from a canonical exponential family. Let $K \subset N^\circ$ be compact. Then $\bar{x}_n$ is uniformly asymptotically normal over $\theta \in K$ with mean $\xi(\theta)$ and covariance matrix $n^{-1}\not{Z}(\theta) = n^{-1}D_2\psi(\theta)$. [Apply Theorem 2.19.]

5.15.2

Consider the setting of 5.15.1: (a) The maximum likelihood

estimators $\hat{\theta}_n$ and $\hat{\xi}_n$ exist with probability approaching 1 as $n \to \infty$ uniformly over $\theta \in K$. (b) They are asymptotically normal uniformly over $\theta \in K$ with means θ and ξ and covariances $n^{-1}\Sigma^{-1}(\theta)$ and $n^{-1}\Sigma(\theta)$, respectively. [(a) $P(\bar{x}_n \notin R)$ converges to 0 (exponentially fast), uniformly on K. (b) if $g(t) = g(t_0) + (h(t_0))'(t - t_0) + o(||t - t_0||)$ then $g(\bar{x}_n)$ is asymptotically normal with mean $\xi(\theta)$ covariance $h'(\xi(\theta))\,\Sigma(\theta)h(\xi(\theta))$, uniformly for $\theta \in K$.]

5.15.3

Let $X_1,\ldots,X_n$ be i.i.d. with distribution p from a differentiable exponential subfamily $\{p_\theta: \theta \in \Theta\}$. Let $K \subset \Theta$ be compact. (a) Then $\hat{\theta}_n$ is uniformly asymptotically normal over $\theta \in K$ with mean θ. (b) For a curved exponential family with $\theta = \theta(t)$ the maximum likelihood estimator $\hat{t}_n$ of t is uniformly asymptotically normal at $\theta(t) \in K$ with mean t. (c) Write the asymptotic variance of $\hat{t}_n$ as a function of $\Sigma(\theta(t))$, $\theta'(t)$, and the statistical curvature at t of the curved exponential family. [See Theorem 5.12, the hint to 5.15.2(b), and Section 3.11. For (c), and for a geometric interpretation of (a) and (b) note that $\sqrt{n}||\hat{\xi}_n - \hat{\hat{\xi}}_n|| \to 0$ in probability where $\hat{\hat{\xi}}_n$ denotes the projection in the inner product $\langle s, t\rangle = s'\,\Sigma^{-1}(\theta)t$ of $\bar{x}_n$ on the tangent line at θ to Θ. If the problem is written in the canonical form of Section 3.11 the asymptotic variance is I.]

5.15.4

Let $\{p_\theta: \theta \in \Theta\}$ be a curved exponential family. Let $\theta' \in N$ but $\theta' \notin \Theta$. Assume (w.l.o.g.) that the family has been written in the canonical form 3.11(1) - (4) with $0 = \hat{\xi}_\Theta(\theta') = \hat{\theta}_\Theta(\xi(\theta'))$. Show $\theta' = (0,\alpha,\ldots,0)$ with $\alpha \le \rho$. Let $X_1,\ldots,X_n$ be i.i.d. observations under θ' from this family and let $\hat{t}$ be the maximum likelihood estimator of t. Show that if $\alpha<\rho$ then t is asymptotically normal about 0 with variance $\sigma_{11}(\theta')\rho^2/(\rho - \alpha)^2$. What happens when $\alpha = \rho$?

CHAPTER 6. THE DUAL TO THE MAXIMUM LIKELIHOOD ESTIMATOR

KULLBACK-LEIBLER INFORMATION (ENTROPY)

Before turning to the dual of the maximum likelihood estimator we define the Kullback-Leibler information, and prove a few of its simple properties. The goal of this detour is to provide a natural probabilistic interpretation for this dual as the minimum entropy expectation parameter.

6.1 Definitions

Suppose F, G are two probability distributions with densities f, g relative to some dominating σ-finite measure ν. The *Kullback-Leibler information* of G at F is

(1) $$K(F, G) = E_F(\ln(f(x)/g(x)))$$

with the convention that $\infty \cdot 0 = 0$, $0/0 = 1$, and $y/0 = \infty$ for $y > 0$. K is also referred to as the *entropy* of G at F.

It can easily be verified that K(F, G) is independent of the choice of dominating measure ν. The existence of K will be established in Lemma 6.2 where it is shown that $0 \leq K \leq \infty$.

In exponential families it is convenient to write

(2) $$K(\theta_0, \theta_1) = K(P_{\theta_0}, P_{\theta_1}) , \quad \theta_0, \theta_1 \in N \quad .$$

For $S \subset N$ let

(3) $$K(S, \theta_1) = \inf\{K(\theta_0, \theta_1): \theta_0 \in S\} \quad ,$$

etc.

$K(\cdot,\cdot)$ as defined in (2) has domain $N \times N$. It is convenient to also transfer this definition to the expectation parameter space. Accordingly, define $\tilde{K}(\xi_0, \xi_1)$ by

$$\tilde{K}(\xi_0, \xi_1) = K(\theta(\xi_0), \theta(\xi_1)) \tag{4}$$

for $(\xi_0, \xi_1) \in \xi(N^\circ) \times \xi(N^\circ)$. If the family is steep this definition is valid on $K^\circ \times K^\circ$.

It is also sometimes convenient to extend the definition of $\tilde{K}(\cdot, \xi_1)$ to all of R^K, by lower semicontinuity. Accordingly, for a minimal steep family, and for $\xi_0 \in \bar{K} - K^\circ$, $\xi_1 \in K^\circ$, define

$$\tilde{K}(\xi_0, \xi_1) = \liminf_{\varepsilon \downarrow 0} \{\tilde{K}(\xi, \xi_1) : \xi \in K^\circ, ||\xi - \xi_0|| < \varepsilon\} \quad . \tag{5}$$

For $\xi \notin \bar{K}$, $\xi_1 \in K^\circ$ define

$$\tilde{K}(\xi, \xi_1) = \infty \quad . \tag{6}$$

It is to be emphasized that this is a formal, analytic extension of the definition. $\tilde{K}(\xi_0, \xi_1)$ for $\xi_0 \notin K^\circ$ does not necessarily have a probabilistic interpretation like (1). (Sections 6.18+ give a probabilistic interpretation of $\tilde{K}$, valid under some auxiliary conditions.)

K is often called the Kullback-Leibler "distance" from θ_0 to θ_1, but it is not a metric in the topological sense. In particular, it is -- in general -- not symmetric. There is, however, one very important special case where K is symmetric and $(K)^{\frac{1}{2}}$ is a metric: the normal location family, $\{P_\theta\} = \{\Phi_{\theta,\Sigma}: \theta \in R^k$, forms a standard exponential family with canonical statistic $\Sigma^{-1}x$ (see Example 1.14), and has

$$K(\theta_0, \theta_1) = (\theta_1 - \theta_0)'\Sigma^{-1}(\theta_1 - \theta_0)/2 \quad . \tag{7}$$

The following proposition has already been mentioned above.

6.2 Proposition

For any two distributions K(F, G) exists and satisfies

$$(1) \qquad 0 \le K(F, G) \le \infty \quad .$$

K(F, G) = 0 if and only if F = G.

Proof.
$$E_F(\ln(f(X)/g(X))) = E_F(-\ln(g(X)/f(X)))$$
$$\ge -\ln E_F(g(X)/f(X))$$
$$= -\ln 1 = 0$$

by Jensen's inequality, with equality if and only if f = g a.e.(ν). ||

For exponential families K has an especially simple and appealing form.

6.3 Proposition

Let $\{p_\theta\}$ be a standard exponential family. If $\theta_0 \in N^\circ$, $\theta_1 \in N$ then

$$(1) \qquad K(\theta_0, \theta_1) = (\theta_0 - \theta_1) \cdot \xi(\theta_0) - (\psi(\theta_0) - \psi(\theta_1))$$
$$= \log (p_{\theta_0}(\xi(\theta_0))/p_{\theta_1}(\xi(\theta_0))) \quad .$$

(*Remark.* Suppose $\{p_\theta\}$ is steep and $\theta_0 \in N - N^\circ$, $\theta_1 \in N^\circ$. Then $K(\theta_0, \theta_1) = \infty = \lim_{\eta_i \to \theta_0} K(\eta, \theta_1)$ for $\{\eta_i\} \subset N^\circ$ by steepness. Since the only sensible interpretation for $(\theta_0 - \theta_1) \cdot \xi(\theta_0)$ is ∞ here, (1) may be considered valid for all $\theta_0 \in N$ for regular or steep families.)

Proof. Note that

$$\ln(p_{\theta_1}(x)/p_{\theta_0}(x)) = (\theta_1 - \theta_0) \cdot x - (\psi(\theta_1) - \psi(\theta_0))$$

and $E_{\theta_0}(X) = \xi(\theta_0)$. ||

6.4 Remark

The second part of 6.3(1) shows how the Kullback-Leibler information is related to maximum likelihood estimation. For $S \subset N$ let

(1) $$K(\theta_0, S) = \inf\{K(\theta_0, \theta_1): \theta_1 \in S\} \quad .$$

Then, by 6.3(1), if $\theta_0 \in N^\circ$

(2) $$K(\theta_0, S) = K(\theta_0, \theta)$$

for $\theta \in S$ if and only if $\theta \in \hat{\theta}_S(\xi(\theta_0))$.

In other words, for steep families, for $\Theta = S$, and for an observation $x \in K^\circ$ the maximum likelihood estimator is the closest point in S to $\theta(x)$ in the Kullback-Leibler sense. (For observations $x \in K - K^\circ$ such an interpretation requires an extension of the definition of K like that to be provided in Sections 6.18+.)

Note also that

(3) $$K(\theta_0, \theta_1) = \ell(\theta_0, \xi(\theta_0)) - \ell(\theta_1, \xi(\theta_0)) \quad .$$

The fact that the quantity on the right is positive (for $\theta_0 \in N^\circ$, $\theta_1 \neq \theta_0$) has already been used in 5.8(3) and 5.12(3).

6.5 Theorem

Let $\{p_\theta\}$ be a standard exponential family. Then $K(\cdot,\cdot)$ is infinitely differentiable on $N^\circ \times N^\circ$. On N°

(1) $$\nabla K(\theta_0, \cdot) = \xi(\cdot) - \xi(\theta_0)$$

(2) $$D_2 K(\theta_0, \cdot) = D_2\psi(\cdot) = \Sigma(\cdot) \,, \quad \theta_0 \in N^\circ$$

If $\{p_\theta\}$ is minimal and steep then on K°

(3) $$\nabla \tilde{K}(\cdot, \xi_1) = \theta(\cdot) - \theta(\xi_1)$$

(4) $$D_2\tilde{K}(\cdot, \xi_1) = \Sigma^{-1}(\theta(\cdot)) , \qquad \xi_1 \in K^\circ \quad .$$

Consequently, given $\xi_1 \in K^\circ$ and $\varepsilon_1 > 0$ there is an $\varepsilon_2 > 0$ such that

(5) $$\tilde{K}(\xi, \xi_1) \geq \varepsilon_2 ||\xi - \xi_1|| \quad \text{whenever} \quad ||\xi - \xi_1|| > \varepsilon_1 \quad .$$

If $S \subset K^\circ$ is compact then a value $\varepsilon_2 > 0$ can be chosen so that (5) is valid uniformly for all $\xi_1 \in S$.

Proof. Formulae (1) - (3) are straightforward from 6.3(1). (Note also that (1), (2) are merely a restatement of 5.3(1), (2).) (4) follows from (3) by the inverse function theorem since $\theta(\cdot) = \xi^{-1}(\cdot)$ and $\nabla\xi(\cdot) = \Sigma(\cdot)$. Formula (5) follows from (3), (4) as did the analogous conclusion 5.3(3), and 5.3(5) of Lemma 5.3 follow from 5.3(1), (2). The asserted uniformity of (5) over $\xi_1 \in S$ is easy to check in that proof. ||

(Note: if p_θ is not minimal 6.5(3) is still valid and 6.5(4) is valid with Σ^{-1} interpreted as a generalized inverse.)

CONVEX DUALITY

6.6 Definition

Let $\phi: R^k \to (-\infty,\infty]$ be convex. The *convex dual* of ϕ is the function $d_\phi: R^k \to [-\infty, \infty]$ defined by

(1) $$d_\phi(x) = \sup\{\ell_\phi(\theta, x): \theta \in R^k\} \quad .$$

(Recall, $\ell_\phi(\theta, x) = \theta \cdot x - \phi(\theta)$.)

We will be interested in the situation when ϕ is regularly strictly convex and steep. (See Definition 5.2.) Then if $x \in R = \xi(N_\phi^\circ)$, $\ell(\cdot, x)$ is strictly concave on N_ϕ and $\nabla\ell(\cdot, x)_{|\theta(x)} = 0$. Thus

(2) $$d_\phi(x) = \ell_\phi(\theta(x), x) \quad \text{for} \quad x \in R = \xi(N_\phi^\circ) \quad .$$

(In such cases, and somewhat more generally, the pair (d_ϕ, R) is called the

Legendre transform of (ϕ, N_ϕ). It is easy to check from (2) and Theorem 6.5 that

(3) $$d_{d_\phi}(\theta) = \phi(\theta) \quad \text{for} \quad \theta \in N^\circ \quad .$$

It can be shown that (3) actually holds for all $\theta \in R^k$, but we do not need this fact in what follows.)

Suppose ψ is the cumulant generating function of a steep exponential family. Then

(4) $$d_\psi(x_0) = \tilde{K}(x_0, x_1) + \theta(x_1) \cdot x_0 - \psi(\theta(x_1)) , \qquad x_0 \in K^\circ \quad .$$

If the coordinate system and dominating measure are chosen so that $\psi(0) = 0 = \xi(0)$ then (4) becomes

(4') $$d_\psi(x_0) = \tilde{K}(x_0, 0) \qquad x \in K^\circ \quad .$$

This provides a probabilistic interpretation for $d(x)$ on K°. It will be seen later that $d(\cdot)$ is the maximal lower semicontinuous extension of $(d(x)\colon\ x \in K^\circ)$ to all of R^k, and (4) is valid for all $x_0 \in R^k$.

Lemmas 6.7 and 6.8 and Theorem 6.9 present some important basic facts about convex duality. They are just the tip of a rich theory. We will not further develop this theory as an abstract unit; although other important features of the theory are implict in results we state elsewhere (e.g. Theorem 5.5). A unified presentation of the theory appears in Rockafeller (1970), and many elements of it are in Barndorff-Nielsen (1978, especially Chapters 5 and 9).

<u>6.7 Lemma</u>

The convex dual d is a lower semicontinuous convex function. Hence, N_d is convex. Suppose ϕ is regularly strictly convex. Then d is strictly convex and twice differentiable on R. On R

(1) $$\nabla d(x) = \theta(x) ,$$

and

(2) $$D_2 d(x) = (D_2 \phi)^{-1} (\theta(x)) .$$

Proof. Since d is the supremum of linear functions it is lower semi-continuous and convex.

For $x \in R$, $d(x) = x \cdot \theta(x) - \psi(\theta(x))$. Hence (1), (2) hold, by the same computation that yielded 6.5(3), (4). d is strictly convex on R since $D_2 d$ is positive definite. (It is possible to also directly establish strict convexity without requiring that ϕ be twice differentiable.) ||

It is now convenient to consider

$$\ell_d(x, \theta) = x \cdot \theta - d(x) .$$

Under the conditions of Lemma 6.7 $\nabla d(x) = \theta(x)$ so that for $\theta \in N°$ $\ell_d(\cdot, \theta)$ is uniquely maximized at the value x for which $\theta(x) = \theta$. This value is precisely $\xi(\theta)$. This interpretation is developed further below, especially in Definition 6.10.

The following equivalent expression for steepness is a fundamental building block in the proof of Theorem 6.9, and has other uses.

6.8 Lemma

Let ϕ be regularly strictly convex. Then ϕ is steep if and only if

(1) $$(\{\theta_i\} \subset N°, \quad \theta_i \to \theta \in N - N°)$$

implies

(2) $$||\nabla\phi(\theta_i)|| \to \infty .$$

Proof. Assume (1) implies (2). Let $\theta_0 \in N°$, $\theta_1 \in N - N°$, $\theta_\rho = \theta_0 + \rho(\theta_1 - \theta_0)$. Then

(3) $$-\ell_d(\xi(\theta_\rho), \theta_0) = d(\xi(\theta_\rho)) - \xi(\theta_\rho) \cdot \theta_0$$

$$= \xi(\theta_\rho) \cdot (\theta_\rho - \theta_0) - \phi(\theta_\rho) \quad .$$

d is strictly convex and twice differentiable on the open set R with $(D_2 d)$ nonsingular on R. Hence

(4) $$\lim_{||x|| \to \infty} \ell_d(x, \theta) = -\infty$$

for every $\theta \in \theta(R) = N°$ by Lemma 5.3(3). Since $||\xi(\theta_\rho)|| \to \infty$, by (2), we have

(5) $$\xi(\theta_\rho) \cdot (\theta_\rho - \theta_0) - \phi(\theta_\rho) = -\ell_d(\xi(\theta_\rho), \theta_0) \to \infty.$$

Since $\theta_1 \in N$, $\lim_{\rho \to 1} \phi(\theta_\rho) = \phi(\theta_1)$ is finite. This implies

(6) $$\xi(\theta_\rho) \cdot (\theta_1 - \theta_0) = \xi(\theta_\rho) \cdot (\theta_\rho - \theta_0)/\rho \to \infty \quad \text{as} \quad \rho \uparrow 1 \quad .$$

By definition, ϕ is steep.

Conversely, suppose there is a sequence satisfying (1) for which (2) fails. The sequence can be chosen so that

$$\sup ||\nabla\phi(\theta_i)|| = B < \infty \quad .$$

This means that $\xi(\theta_i) = \nabla\phi(\theta_i)$, $i=1,\ldots$ is a bounded sequence, thus, without loss of generality, the original sequence $\{\theta_i\}$ can be assumed to have been chosen to satisfy $\xi(\theta_i) \to x^*$.

Hence, for any $\theta' \in R^k$

$$(7) \qquad \theta \cdot x^* - \phi(\theta) = \lim (\theta_i \cdot \xi(\theta_i) - \phi(\theta_i))$$

$$\geq \limsup (\theta' \cdot \xi(\theta_i) - \phi(\theta'))$$

$$= \theta' \cdot x^* - \phi(\theta') \quad .$$

It follows that

$$(8) \qquad d(x^*) = \theta \cdot x^* - \phi(\theta) < \infty \quad .$$

This means that $\theta \notin N^\circ$ satisfies $\theta \in \hat{\theta}(x^*)$. By Theorem 5.5 this is impossible if ϕ is steep. Hence ϕ is not steep. ||

Proof of Proposition 3.3. It is now easy to prove the converse assertion in Proposition 3.3, namely that a minimal exponential family satisfying

$$(9) \qquad E_\theta(||x||) = \infty \quad \text{for} \quad \theta \in N - N^\circ$$

is steep.

By Fatou's lemma if $\{\theta_i\}$ satisfies (1) then

$$\lim ||\nabla\psi(\theta_i)|| = \lim ||E_{\theta_i}(x)|| \geq \lim E_{\theta_i}(||x||) = \infty \quad .$$

Hence (2) is satisfied. Thus ψ is steep, which is the desired result. ||

6.9 Theorem

Assume ϕ is steep and regularly strictly convex. Then d_ϕ is also, and

$$(1) \qquad N^\circ_{d_\phi} = R_\phi = \xi(N^\circ) \quad .$$

Proof. Let $x_0 \in R$, $v \in R^k$. Let $\rho_v = \inf \{\rho > 0: x_0 + \rho v \notin R\}$. Note that $\rho_v > 0$ since R is open. Assume $\rho_v < \infty$ and let $x_1 = x_0 + \rho_v v$ and $x_\rho = x_0 + \rho(x_1 - x_0)$. Note that $x_1 \notin R$.

Suppose it were true that

(2) $$\liminf_{\rho \uparrow 1} ||\theta(x_\rho)|| < \infty \quad .$$

Then there would be a sequence $\rho_i \uparrow 1$ with $\theta(x_{\rho_i}) \to \theta^*$, say. $\theta^* \notin N^\circ$ since $x_1 \notin R = \xi(N^\circ)$. But then, since ϕ is steep, this would imply

$$||x_{\rho_i}|| = ||\xi(\theta(x_{\rho_i}))|| \to \infty$$

by Lemma 6.8, which is a contradiction since $x_{\rho_i} \to x_1$. Hence (2) is false; so that actually

(3) $$\lim_{\rho \uparrow 1} ||\theta(x_\rho)|| = \infty \ .$$

The argument in the first part of the proof of Lemma 6.8 applies to yield the dual to 6.8(6), namely

(4) $$\theta(x_\rho) \cdot (x_1 - x_0) \to \infty \quad \text{as} \quad \rho \uparrow 1 \quad .$$

(Technically, the lemma as stated cannot be directly quoted since we have not yet established that $R = N_d$ so that d is regularly strictly convex. But, d has the desired convexity and differentiability properties on $R \subset N_d$ by Lemma 6.7. It is then easy to check that the first part of Lemma 6.8 indeed applies since $\{x_{\rho_i}\} \subset R$ and yields (4) as the dual of 6.8 (6).)

d is therefore a convex function with

(5) $$\frac{d}{d\rho} d(x_0 + \rho(x_1 - x_0)) \to \infty \quad \text{as} \quad \rho \uparrow 1 \quad .$$

This implies that

(6) $$d(x_0 + \rho(x_1 - x_0)) = \infty \quad \text{for} \quad \rho > 1 \ .$$

Since the above argument applies for all $v \in R^k$, it yields that

(7) $$d(x) = \infty \quad \text{for} \quad x \notin \bar{R} \quad .$$

Thus $\bar{R} \supset N_d$. This yields (1) since, also, $R \subset N_d$ because $d(x) = \theta(x) \cdot x - \phi(\theta(x)) < \infty$ on R.

It now follows that d is regularly strictly convex since it has the desired smoothness properties, etc., on $R = N_d^\circ$ by Lemma 6.7. And, finally, d is steep since (5) applies to any $x_0 \in R$, $x_1 \in \bar{R} - R$. ||

Remark. Since d is convex, lower semicontinuous, and $d(x) = \infty$ for $x \notin \bar{R}$ it must be that $d(\cdot)$ on R^k is the maximal lower semicontinuous extension of $d(x)$: $x \in R$ $(= K^\circ)$ to all of R^k. That is, for $x_1 \in \bar{R} - R$

$$d(x_1) = \liminf_{\varepsilon \downarrow 0} \{d(x): x \in R, ||x - x_1|| < \varepsilon\} .$$

It follows that if $\{p_\theta\}$ is a steep exponential family. The relation 6.6(4) between $d(x_0)$ and $\tilde{K}(x_0, x_1)$ is valid for all $x_0 \in R^k$, $x_1 \in K^\circ$.

MINIMUM ENTROPY PARAMETER

The path has been prepared for the definition of the dual to maximum likelihood estimation, and for the basic existence and construction theorems.

6.10 Definition

Let $d: R^k \to (-\infty, \infty]$ be convex and lower semicontinuous. Let $S \subset R^k$. Define

$$(1) \qquad \tilde{\xi}_S(\theta) = \{\xi \in S: \ell_d(\xi, \theta) = \ell_d(S, \theta) = \inf \{\ell_d(x, \theta): x \in S\}\} .$$

Obviously $\tilde{\xi}_S$ is related to ℓ_d in the same fashion as $\hat{\theta}$, the maximum likelihood estimator for an exponential family, is related to the log likelihood function ℓ_ψ. (It would therefore seem logical to adopt the notation $\hat{\xi}_S$ rather than $\tilde{\xi}_S$. However for reasons of convenience and tradition we wish to reserve the notation $\hat{\xi}_S$ for the set of maximum likelihood estimates of expectation parameters. That is, $\hat{\xi}_S(x) = \xi(\hat{\theta}_S(x))$.)

The function $\tilde{\xi}_S$ has been given a variety of fairly inconvenient appelations. For example, values in $\tilde{\xi}_S(\theta)$ can be called *minimum entropy* (expectation) *parameters* relative to the set $S \subset K°$. Barndorff-Nielsen (1978) refers to values $\tilde{\theta}_S(x) = \theta(\tilde{\xi}_S(\theta(x)))$, $x \in K°$, as *maximum likelihood predictors*. (Note however that $\tilde{\xi}_S(\theta) \cap (K - K°) \neq \phi$ is possible even if $\{p_\theta\}$ is regular as long as S is not convex (see Theorem 6.13). Hence values in $\tilde{\xi}$ need not always be expectation parameters.)

Another interpretation is provided by the Kullback-Leibler information. Consider a steep minimal exponential family. If $\tilde{\xi} \in \tilde{\xi}_S(\theta) \cap K°$ then

$$\tilde{K}(\tilde{\xi}, \xi(\theta)) = \inf \{\tilde{K}(x, \xi(\theta)): x \in S \cap K°\} .$$

Thus, $\tilde{\theta} \in \theta(\tilde{\xi}_S(\theta_1))$ is a parameter in $\theta(S)$ whose Kullback-Leibler distance to θ_1 is a minimum over all parameters in $\theta(S)$.

Suppose $\{p_\theta\}$ is a minimal, steep standard exponential family. Then Theorem 6.9 establishes that d_ψ is steep and regularly strictly convex with $R = \xi(N°) = K°$. Consequently $\tilde{\xi}$ possesses the properties established for $\hat{\theta}$ in Chapter 5. The main properties are formally stated below; their proofs consist only of reference to the appropriate results in Chapter 5.

Convention. In the following statements $\{p_\theta\}$ is a minimal steep standard exponential family. Note that $R = K^\circ \subset N_d \subset K$.

<u>6.11 Theorem</u>

If $\theta \in N°$ then

(1) $$\tilde{\xi}_N(\theta) = \{\xi(\theta)\} \subset K° .$$

If $\theta \in N - N°$ then $\tilde{\xi}_N(\theta)$ is empty.

Proof. This is the dual statement to Theorem 5.5. ||

Note that

(2) $$\theta(\tilde{\xi}_N(\theta(x))) = \hat{\theta}_N(x) , \quad \text{etc.}$$

In other words, for a full exponential family the maximum likelihood predictor is the same as the maximum likelihood estimator. However (2) does not extend to non-full families.

6.12 Theorem

Let $S \subset N_d$ be a non-empty, relatively closed subset of N_d. Suppose $\theta \in N^\circ$. Then $\tilde{\xi}(\theta)$ is non-empty.

Suppose $\theta \in N - N^\circ$ and there are values $\theta_i \in N^\circ$, $i=1,\ldots,I$ and constants $\beta_i < \infty$ such that

(1) $$S \subset \bigcup_{i=1}^{I} H^-(\theta - \theta_i, \beta_i) \quad .$$

Then $\tilde{\xi}(\theta)$ is non-empty.

For any $\tilde{\xi} \in \tilde{\xi}_S(\theta) \cap K^\circ$

(2) $$\theta - \theta(\tilde{\xi}) \in \nabla_S(\tilde{\xi}) \quad .$$

Proof. Invoke Theorem 5.7 and Theorem 5.12. ||

6.13 Theorem

Suppose $S \cap N_d$ is a relatively closed convex subset of N_d with $S \cap K^\circ$ non-empty. Then $\tilde{\xi}_S(\theta)$ is non-empty if and only if $\theta \in N^\circ$ or $\theta \in \bar{N} - N^\circ$ and

(1) $$S \subset H^-(\theta - \theta_1, \beta_1)$$

for some $\theta_1 \in N^\circ$, $\beta_1 \in R$.

If $\tilde{\xi}_S(\theta)$ is non-empty then it consists of the unique point $\tilde{\xi} \in S \cap K^\circ$ satisfying

(2) $$(\theta - \theta(\tilde{\xi})) \cdot (\tilde{\xi} - \xi) \geq 0 \qquad \forall \quad \xi \in S \quad .$$

Proof. Invoke Theorem 5.8. ||

6.14 Construction

Theorems 6.12(2) and 6.13 have a geometrical interpretation which looks exactly like that of their counterparts in Chapter 5. For example, suppose $S = H \cap K$ with H the hyperplane $H(a, \alpha)$, and $H \cap K^\circ$ is non-empty. Then in order to find $\tilde{\xi}_S(\theta)$ one need only search for the unique point $\xi^* \in H$ for which $\theta - \theta(\xi^*) = \rho a$ for some $\rho \in R$. The process can be pictured from two different perspectives. Both of these are shown in Figure 6.14(1).

(i) One may proceed from $\xi(\theta)$ along the curve $\{\xi(\theta + \rho a): \rho \in R\}$ until the unique point at which $\xi(\theta + \rho a) \in H$.

(ii) Alternatively one may map $S \cap K^\circ$ back into Θ as $\theta(S \cap K^\circ)$ and then proceed along the line $\{\theta + \rho a: \rho \in R\}$ until the unique point at which $\theta + \rho a \in \theta(S \cap K^\circ)$.

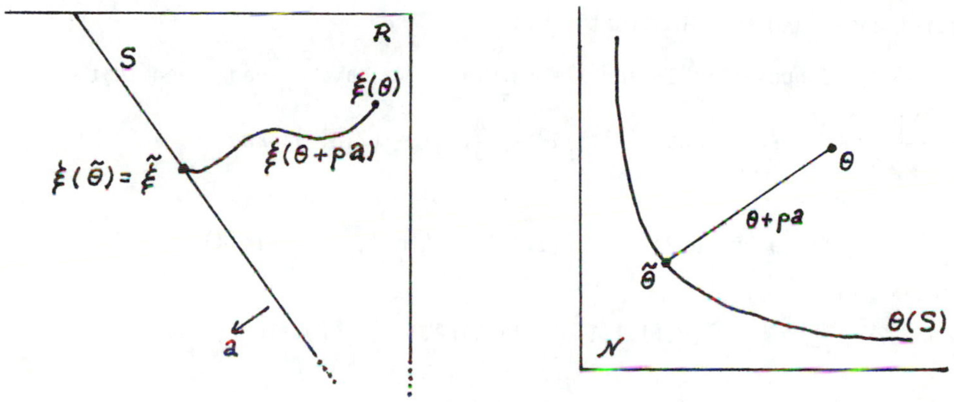

Figure 6.14(1): Construction of $\tilde{\xi}_S(\theta)$ when $S = H(a, \alpha) \cap K$

There is an important statistical difference between the situation pictured here and the dual situation displayed in 5.9.

In Construction 5.9 $\Theta = H \cap N$ and the problem considered was to find $\hat{\theta}_\Theta$. In that case one could proceed via the geometrical dual to Figure 6.14(1). See Figures 5.9(1) and 5.9(2). However, one could also reduce by sufficiency to a minimal exponential family with parameter space Θ. $\hat{\theta}_\Theta$ could then be found by applying Theorem 5.5 to this minimal family. A corresponding statistical interpretation is not available for the dual problem of finding $\tilde{\xi}_{H\cap K}$.

Furthermore, if $\Theta = H \cap N$ and $S = \xi(\Theta)$ the maximum likelihood predictor relative to S cannot legally be found by first reducing by sufficiency. This very undesirable property of a statistical estimator is displayed in the following example.

6.15 Example

Consider the Hardy-Weinberg problem discussed earlier in Examples 1.8 and 5.10. Let $S = \xi(\Theta)$ and consider the problem of finding $\tilde{\xi}_S$. Rather than provide a general formula for $\tilde{\xi}$ (a messy exercise) we discuss a special case, and some implications.

Suppose N = 18 and x = (3,6,9). We have already seen that $\hat{p} = \frac{2x_1 + x_2}{2N} = \frac{1}{3}$. Thus $\hat{\xi}(x) = 18(\frac{1}{9}, \frac{4}{9}, \frac{4}{9}) = (2,8,8)$, and

$$(1) \qquad \theta(\hat{\xi}(x)) = \hat{\theta}(x) = \{\rho(1,1,1) + (\ln 1, \ln 4, \ln 4)\}$$

$$= \{\beta_1(1,1,1) - (\ln 2)(2,1,0) + (0, \ln 2, 0)\} \subset \Theta \quad .$$

Note also that

$$(2) \qquad \theta(x) = \{\rho(1,1,1) + (\ln 1, \ln 2, \ln 3)\} \quad .$$

Of course $\theta(x) \cap \Theta = \phi$.

Since $\xi(p) = (p^2, 2pq, q^2) = (p^2, 2p(1-p), (1-p)^2)$ the tangent space to $S = \{\xi(p): \ 0 < p < 1\}$ can be found by taking $\frac{d}{dp}\,\xi(P)$. Evaluated at $\hat{p} = \frac{1}{3}$ this tangent space, T, is spanned by the vector

$$\tau = (2\hat{p}, \ 2 - 4\hat{p}, \ -2 + 2\hat{p})$$

$$= (\tfrac{2}{3}, \ \tfrac{2}{3}, \ -\tfrac{4}{3}) \quad .$$

By definition $\nabla_S(\hat{\xi}) = \{v: \ v \cdot \tau = 0\}$.

Now, from (1) and (2)

$$\theta(x) - \theta(\hat{\xi}) = \{\rho'(1,1,1) + (0, \ln 2 - \ln 4, \ln 3 - \ln 4): \rho' \in R\} .$$

Thus

(3) $$(\theta(x) - \theta(\hat{\xi})) \cdot \tau = (2/3) \ln (1/2) - (4/3) \ln (3/4) \neq 0 .$$

The implication of (3) is that $\theta(x) - \theta(\tilde{\xi}) \notin \nabla_S(\hat{\xi})$. It follows from Theorem 6.12(2) that

(4) $$\tilde{\theta}(x) \cap \hat{\theta}(x) = \phi ,$$

or, in other words,

(4') $$\tilde{\xi}(x) \neq \hat{\xi}(x) .$$

Finally, suppose instead that the sample point is $x^* = (2,8,8)$. Note that $x^* = \hat{\xi}(x)$ with $x = (3,6,9)$, as above. In this case $\hat{\xi}(x^*) = x^*$ and hence

(5') $$\tilde{\xi}(x^*) = \hat{\xi}(x^*) = x^*$$

and

(5) $$\tilde{\theta}(x^*) = \hat{\theta}(x^*) = \theta(x^*) .$$

Recall from the discussion in Example 5.10 that, over the domain K°, $\hat{\xi}(x)$ coincides with the minimal sufficient statistic. Thus, from (4) and (5) (or (4') and (5')) it can be seen that here the "estimator" $\tilde{\theta}(x) = \theta(\tilde{\xi}(\theta(x)))$ *is not a function of the minimal sufficient statistic*. This is a very undesirable property for a statistical estimator. Indeed, we emphasize, the primary statistical use of $\tilde{\theta}$ does not lie in its use as a statistical estimator, but rather in its use in the theory of large deviations. See, for example, 7.5 and Exercises 7.5.1 - 7.5.6.

ENTROPY

6.16 Discussion

In statistical mechanics and elsewhere the term entropy appears and has a definition whose connection with the quantity $K(\theta_0, \theta_1)$ for exponential families is not at first obvious. See Ellis (1984a; 1984b).

Let F be a probability distribution on R^k. Let $x \in R^k$ and define the *entropy of x under* F as

(1) $$\mathcal{E}_F(x) = \inf \{K(G, F): E_G(X) = x\} \quad .$$

There is, as yet, no exponential family apparent in this definition. However, there is indeed an intimate connection between ξ and $\tilde{K}$, as revealed in the following theorem. The theorem is proved only for the case where F satisfies certain mild assumptions and $x \in \mathcal{K}_F^\circ$ or $x \notin \mathcal{K}_F$. We leave it to the reader to develop the appropriate results when F does not satisfy these assumptions. The situation where $x \in \mathcal{K} - \mathcal{K}^\circ$ can sometimes be treated using the methods at the end of this chapter.

6.17 Theorem

Suppose the exponential family generated by F is a steep minimal family with $0 \in \operatorname{int} \mathcal{N}$. Let $\xi_0 = \xi(0) = E_F(X)$. Let $\tilde{K}$ denote the usual Kullback-Leibler function, 6.1(4), for this exponential family. Then

(1) $$\mathcal{E}_F(y) = \tilde{K}(y, \xi_0)$$

if $y \in \mathcal{K}^\circ$. If $y \notin \mathcal{K}$

(2) $$\infty = \mathcal{E}_F(y) = K(y, \xi_0) \quad .$$

Proof. Suppose $y \in \mathcal{K}^\circ$, it is obviously true that

(3) $$\mathcal{E}_F(y) \le K(y, \xi_0)$$

since the distribution $G(dx) = p_{\theta(y)}(x)F(dx) = P_{\theta(y)}(dx)$ satisfies $E_G(X) = y$

and $K(G, F) = \tilde{K}(y, \xi_0)$. Suppose $K(G, F) < \infty$ and

$$E_G(X) = y = E_{P_{\theta(y)}}(X) \quad . \tag{4}$$

It must be that $G \ll F$, for otherwise $K(G, F) = \infty$. Let $g = \frac{dG}{dF}$, and $p = p_{\theta(y)}$. Then

$$\begin{aligned} K(G, F) - K(P_{\theta(y)}, F) &= \int [g(x) \ln g(x) - p(x) \ln p(x)] F(dx) \\ &= \int g(x)(\ln g(x) - \ln p(x))F(dx) \\ &\quad + \int (g(x) - p(x))(\ln p(x))F(dx) \\ &= K(G, P_{\theta(y)}) \geq 0 \end{aligned} \tag{5}$$

since $\int (g(x) - p(x))(\ln p(x))F(dx) = \int (g(x) - p(x))(\theta \cdot x - \psi(\theta))F(dx) = 0$ by (4). It follows from (3) and (5) that (1) holds. (Also, note that $G = F_{\theta(y)}$ is the unique distribution satisfying (4) and yielding $K(G, F) = E_F(y)$.)

If $y \notin K$ then $E_G(X) = y$ implies $G \not\ll F$ and hence $K(G, F) = \infty = \tilde{K}(y, \xi_0)$. ||

AGGREGATE EXPONENTIAL FAMILIES

If $\{p_\theta\}$ is a full canonical exponential family and $x \in \partial K$ then $\hat{\theta}(x) = \phi$. (See Theorem 5.5.) If $\nu(\partial K) > 0$ then this means that with positive probability the maximum likelihood estimator fails to exist. This occurs most commonly when ν has countable support. In most such cases the family of distributions $\{p_\theta : \theta \in N\}$ can be augmented in a natural way so that the maximum likelihood estimator is always defined over this new, larger family of distributions. The augmented family will be called an aggregate exponential family.

Aggregate exponential families can also be satisfactorily defined in a few special cases where ν does not have countable support, but $\nu(\partial K) > 0$ nevertheless. However, such situations are rare in applications and the general theory involves difficulties not present in the countable case; hence we do not treat such situations below. For similar reasons of convenience we avoid non-regular exponential families.

Special cases of the theory are extremely familiar -- for example the aggregate family of binomial distributions, which is just $\mathcal{B}(n, p)$, $0 \le p \le 1$. The general theory for the case where ν has finite support appears in Barndorff-Nielsen (1978, p.154-158), along with some observations about generalizations.

6.18 Definitions

Let ν be a measure concentrated on the countable subset $\mathcal{X} = \{x_1, x_2, \ldots\} \subset R^k$. Thus

$$(1) \qquad \nu(\{x_i\}) > 0 \quad i=1,2,\ldots , \quad \nu(\mathcal{X}^c) = 0 .$$

Consider the closed convex set $K = K_\nu$. The *faces* of K_ν are the non-empty sets of the form

$$(2) \qquad F = K \cap H(v, \alpha) \quad \text{where} \quad K \subset \bar{H}^-(v, \alpha) .$$

By convention the set K is itself a face of K (corresponding to $v = 0$, $\alpha = 0$).

A face, F, is itself a closed convex subset, which has dimension s, $0 \le s \le k$. (Only the face $F = K$ can have dimension k.) The *relative interior* of F, denoted ri(F) is the interior of F considered as a subset of R^s. An analytic characterization of ri(F) is that $x \in$ ri(F) if $x \in F$ and if for every hyperplane $H \in R^k$ such that $x \in H$ but $F \not\subset H$ then both $F \cap H^+ \ne \phi$, and $F \cap H^- \ne \phi$.

Let F be a face of K. If $\nu(F) > 0$ then the restriction of ν to F, $\nu_{|F}$ is uniquely defined and non-zero. We use the notation $K_{|F} = K_{\nu_{|F}}$. Note that while it is usually true that $K_{|F} = F$ this need not always be the case.

See Exercise 6.18.1.

The first main theorem involves the following structural assumption on X:

(3) For every $\xi \in X$ there is a face F of K such that $K_{|F} = F$ and $\xi \in \text{ri}(F)$.

If X is finite then (3) is clearly satisfied. Another important case where (3) is satisfied is when $X = \{0,1,\ldots\}^k$, as for example when $X_1,\ldots,X_k$ are independent Poisson or independent negative binomial variables. Assumption 6.22(1) provides an easily verified structural condition which implies (3).

6.19 Definition (Aggregate family)

Let X and ν be as in 6.18. Let $\{p_\theta\}$ be the canonical exponential family of densities generated by ν. Assume the family is regular. As shown in Chapter 3 this family can be reparametrized by the expectation parameter $\xi = \xi(\theta)$. Let

(1) $$q_{\xi(\theta)}(x) = p_\theta(x) \qquad \theta \in N \quad .$$

Then, $\{q_\xi: \ \xi \in K^\circ\} = \{p_\theta: \ \theta \in N\}$.

Now, for each face, F, of K with $\nu(F) > 0$ let $\psi_{|F} = \psi_{\nu_{|F}}$ and define the family of densities

$$p_{\theta|F}(x) = \begin{cases} \exp(\theta \cdot x - \psi_{|F}(\theta)) & x \in F \\ 0 & x \notin F \end{cases}$$

relative to the measure ν. This is an exponential family relative to the measure $\nu_{|F}$. Assume this family is regular. Let $\xi_{|F}$ denote its expectation parameter, and let

(2) $$q_{\xi(\theta)|F}(x) = p_{\theta|F}(x) \quad .$$

Thus ξ ranges over the set ri $K_{|F}$ as θ ranges over $N_{|F} = N_{\nu_{|F}}$. Note that the family $\{p_{\theta|F}\colon\ \theta \in N_{|F}\}$ is not minimal. Hence the map $\theta \to \xi_{|F}(\theta)$ is not 1 - 1. However, $q_{\xi_1|F} = q_{\xi_2|F}$ if and only if $\xi_1 = \xi_2$, by virtue of Theorems 1.9 and 3.6.

Let

(3) $$\mathcal{F} = \{x\colon\ \exists \text{ face } F \text{ of } K \ni \nu_{|F} \neq 0 \text{ and } x \in \text{ri}(F)\} .$$

Lemma 6.20, below, establishes that for each $\xi \in \mathcal{F}$ there is a unique F such that $\xi \in \text{ri}(F)$ and a unique density $q_{\xi|F}$ corresponding to the pair ξ, F. This density has

(4) $$E_{q_{\xi|F}}(X) = \xi .$$

We denote this density as q_ξ. The *aggregate family* of densities generated by ν with parameter space $\mathcal{F}$ is the family

(5) $$\{q_\xi\colon\ \xi \in \mathcal{F}\} .$$

Note that

(6) $$P_\xi(X) = 1 \qquad \forall\ \xi \in \mathcal{F} .$$

6.20 Lemma

Make the assumptions in 6.18 and 6.19. Then for each $\xi \in \mathcal{F}$ there is a unique F such that $\xi \in \text{ri}(F)$. The density $q_\xi = q_{\xi|F}$ satisfies 6.19(4). It is, in fact, the unique density of the form $q_{\xi'|F'}$ having expectation ξ.

Proof. Suppose $\xi \in \text{ri}(F)$ and also $\xi \in F' = H(v', \alpha') \cap K$ where $K \subset \bar{H}^-(v', \alpha')$. Then either (i) $F \subset H(v', \alpha')$ or (ii) $F \cap H^+(v', \alpha') \neq \phi$ and $F \cap H^-(v', \alpha') \neq \phi$. In case (ii) $H(v', \alpha')$ is not a supporting hyperplane, a contradiction. Hence (i) holds, and so $F' \supset F$. Reversing the roles of F, F' in the above now shows that $\xi \in \text{ri}(F)$ and $\xi \in \text{ri}(F')$ implies F = F'.

By Theorem 3.6, $\{E_{q_{\xi(\theta)|F}}(x)\colon\ \theta \in N_{\nu_{|F}}\} = \text{ri}(K_{|F}) = \text{ri}(F)$ by 6.18(3) since $\nu_{|F}$ generates a regular family. Thus $q_{\xi|F}$ satisfying 6.19(4) exists.

For every $\xi \in X$ the preceding shows that $\xi = E_{q_\xi}(X) \in ri(F)$ where F is the unique face of K with $\xi \in ri(F)$. Hence $\xi = E_{q_{\xi|F}}(X) = E_{q_{\xi'|F'}}(X)$ implies $F = F'$, and thus, as previously noted, implies $q_\xi = q_{\xi'}$. ||

Assumption 6.18(3) guarantees that $F \supset X$. If the conclusion of 6.18(3) holds for all $\xi \in$ conhull X then $F =$ conhull X. Otherwise it may occur that $F \subsetneq$ conhull X. Exercise 6.20.1 sketches an example. If Assumption 6.22(1) is satisfied then

$$F = \text{conhull } X = K \quad . \tag{1}$$

Here is the first main theorem providing the extension of Theorem 5.5.

6.21 Theorem

Make the assumptions in 6.18 and 6.19. Then for $x \in F \supset X$ the maximum likelihood estimator, $\hat{\xi}(x)$, is uniquely determined by the trivial equation

$$\hat{\xi}(x) = x \quad . \tag{1}$$

Proof. Let $x \in ri(F)$ for some face $F = H(v, \alpha) \cap K$ of K. If $\xi' \in ri(F')$ and $x \notin F'$ then $q_{\xi'}(x) = 0$.

Now suppose $\xi' \in ri(F')$, $x \in F'$, but $F' \neq F$. It follows (as in Lemma 6.20) that $F' \supset F$. The argument now takes place in F'. Hence we can assume for convenience, and without loss of generality, that $F' = R^k \cap K$ and $\xi' \in K^\circ$. We may further assume that $x = 0$, $K \subset \bar{H}^-(e_1, 0)$, and $0 \in ri(F)$ with $F = H(e_1, 0) \cap K$. Then, $\xi' = \xi(\theta')$ for some $\theta' \in N^\circ \subset R^k$. Let $\theta_\rho = \theta' + \rho e_1$, $\rho > 0$. Then

$$q_{\xi(\theta_\rho)}(0) = \exp(-\psi(\theta_\rho)) \tag{2}$$

and

(3) $$e^{\psi(\theta_\rho)} = \int_{x_1<0} e^{\theta'\cdot x+\rho x_1}\nu(dx) + \int_{x_1=0} e^{\theta'\cdot x}\,\nu(dx)$$

$$\downarrow \int_{x_1=0} e^{\theta'\cdot x}\,\nu(dx) = \psi_{|F}(\theta')$$

by the monotone convergence theorem and the definition of $\psi_{|F}$. It follows from (2) and (3) that

(4) $$q_{\xi'}(0) < q_{\xi(\theta_\rho)}(0) < q_{\xi''|F}(0) , \quad 0 < \rho < \infty ,$$

where ξ'' is the unique point in ri(F) defined by $\xi'' = \xi_{|F}(\theta')$.

Finally, if $\xi''' \in$ ri(F) then applying Theorem 5.5 to the measure $\nu_{|F}$ yields

(5) $$q_{\xi'''|F}(0) \leq q_{0|F}(0)$$

with equality only if $\xi''' = 0$. Combining (4), (5), and the first comment in the proof yields

(6) $$\hat{\xi}(0) = 0 .$$

This verifies (1) when $\xi = 0$, and completes the proof. ||

Remark. As noted in the remark preceding the theorem it is usually true that $F \supset$ conhull X. Assume so and assume the hypotheses of the theorem. Let $X_1,\ldots,X_n$ be i.i.d. random variables with density q_ξ, $\xi \in F$. As usual, let $\bar{X}_n = \sum_{i=1}^{n} X_i/n$. Then $\bar{X}_n \in$ conhull $X \subset F$ with probability one. The family of distributions of the sufficient statistic $\bar{X}_n$ is then also an aggregate family fitting the specifications of the theorem. Hence the maximum likelihood estimator of $\xi \in F$ based on $X_1,\ldots,X_n$ satisfies the trivial equation

(6) $$\hat{\xi}(X_1,\ldots,X_n) = \bar{X}_n .$$

The preceding theorem yields the existence of maximum likelihood estimates when the parameter space is F. In order to guarantee existence of these estimates when the parameter space is a proper closed subset of K it suffices to establish continuity in ξ of $q_\xi(x)$, $x \in X$. This continuity is useful for other purposes as well. Somewhat unfortunately, the assumptions of Theorem 6.21 do not imply that $q_\xi(x)$ is continuous in ξ (see Exercises 6.23.5-6) and the following theorems demand stronger assumptions. Sufficient assumptions are described below.

There is a further, aesthetic, reason for wanting to know that $q_\xi(x)$ is continuous in ξ. The definition given in 6.19 of the aggregate family $\{q_\xi(x): \theta \in F\}$ is structurally natural. But there is also an analytically natural definition for the family of distributions generated from $\{p_\theta: \theta \in N\}$ -- namely, the set of all probability distributions on X which are limits of sequences of distributions in $\{p_\theta\}$. These two definitions coincide when $q_\xi(x)$ is continuous in ξ.

6.22 Assumptions

K is called a polyhedral convex set if it can be written as the intersection of a finite number of half spaces (see Rockafellar (1970)). Assume that K is a polyhedral convex set and that for every one of the finite number of faces, F, of K

$$F = K_{|F} \quad . \tag{1}$$

As previously noted in 6.20(1), this implies $F = K = \text{conhull } X$.

For any convex set $S \in R^k$ define the centered span of S to be the subspace spanned by vectors of the form $x - y$, $x,y \in X$. Denote this subspace by csp S. Note that if $x_0 \in \text{ri } S$ then

$$\text{csp } S = \text{span}\{x - x_0: x \in S\} \quad . \tag{2}$$

Assume that for every face F of K

$$\text{Proj}_{\text{csp } F}\, N = \text{Proj}_{\text{csp } F}(N_{|F}) \quad . \tag{3}$$

Note that if X is finite then (1) is satisfied, and (3) is trivially satisfied since $N_{|F} = R^k$ for all faces F (including $F = K$). If ν is a product measure then (1) and (3) are again satisfied. See Exercise 6.22.2.

6.23 Theorem

Make the assumptions in 6.18, 6.19, and 6.22. Then for every $x \in K$, $q_\xi(x)$ is continuous for $\xi \in K$.

Proof. The proof involves an induction on the dimension, k. For $k = 1$ the result is nearly obvious. Suppose $\xi_0 \in \partial K$. Without loss of generality assume $K \subset (-\infty, \xi_0]$. Then $\xi_i \to \xi_0$ with $\xi_i \neq \xi_0$, $i=1,\ldots$ implies $\xi_i = \xi(\theta_i)$, $\theta_i \to N$, and $\theta_i \to \infty$. It follows that $q_{\xi_i}(\xi_0) = p_{\theta_i}(\xi_0) \to \nu(\{\xi_0\})^{-1} = q_{\xi_0}(\xi_0)$, and for $x \neq \xi_0$, $q_{\xi_i}(x) \to 0 = q_{\xi_0}(x)$.

For arbitrary k, including k=1, if $\xi_0 \in K^\circ$ then $q_\xi(x) = p_{\theta(\xi)}(x)$ is continous on a neighborhood of ξ_0. This completes the proof for $k = 1$.

We now turn to the case $k \geq 2$. We need to prove continuity of q_ξ at $\xi_0 \in \partial K$. Let $\xi_i \to \xi_0$. We need consider only the case where $\{\xi_i\} \subset F$ with F some face of K, since K has only a finite number of faces. If this F is a proper face of K then $q_{\xi_i} \to q_{\xi_0}$ by the induction hypothesis. Hence we need consider only the case where each $\xi_i = \xi(\theta_i)$, $\theta_i \in N$.

There is a unique face F_0 of K such that $\xi_0 \in \text{ri } F_0 = \text{ri } K_{|F_0}$. Without loss of generality assume $\xi_0 = 0$, $K \subset \bar{H}^-(e_1, 0)$, $-\sigma e_1 \in K^\circ$ for some $\sigma > 0$, $F_0 = H(e_1, 0) \cap K$ and $\text{csp } F_0 = \{w \in R^k\colon\ w = (0, \omega),\ \omega \in R^s\}$, $(0 \leq s \leq k-1)$. Let $S = \text{csp } F_0$. For $w \in R^k$ write $w' = (w'_{(1)}, w'_{(2)})$ with $w_{(2)} \in R^s$. Further, assume $0 \in N_{|F_0}$, $\psi_{|F_0}(0) = 0$, $\xi_{|F_0}(0) = 0$. Note that $\psi_{|F_0}(\theta)$ is a function of $\theta_{(2)}$, and so we will write $\psi_{|F_0}(\theta_{(2)})$, where convenient.

We have already assumed $0 \in N_{|F_0}$. Hence $\{\theta \in S\colon\ ||\theta|| \leq \delta_0\} \subset N_{|F_0}$ for some $\delta_0 > 0$. It then follows from 6.22(3) that for each such θ there is a $\sigma(\theta)$, say, such that $\theta + \sigma e_1 \in N$, $\theta \geq \sigma(\theta)$. Since $\{\theta \in S\colon\ ||\theta|| \leq \delta_0\}$ is compact, with $\{\theta \in S: \ ||\theta|| \le \delta_0, \ \theta + \sigma e_1 \in N\}$ as a relatively open subset, there must, further, exists a $\sigma_0 \ge 0$ such that $\theta + \sigma e_1 \in N$ for all $\sigma \ge \sigma_0$, $\theta \in S$, $||\theta|| \le \delta_0$.

For $\delta \le \delta_0$, $\sigma \ge \sigma_0$ define

$$(1) \qquad Q = Q(\sigma, \delta) = \{\theta \in R^k: \ ||\theta_{(2)}|| \le \delta,$$

$$\theta_{(1)} \cdot x_{(1)} \le -\sigma||x_{(1)}|| \quad \forall \quad x \in K\} \quad .$$

Note that $\theta_{(1)} \cdot x_{(1)} - \sigma_0 e_1 \cdot x_{(1)} \le (-\sigma + \sigma_0)||x_{(1)}|| \le 0, \quad \forall \ x \ \in K$.

Hence for $\theta \in Q$

$$\lambda(\theta) \le \lambda(\sigma_0 \, e_1) < \infty$$

as in 6.21(4). It follows that $Q \subset N$.

Now assume for convenience, and without loss of generality, that $\sigma_0 = 0$. Then for $\theta \in Q$

$$(2) \qquad \lambda(\theta) = \int e^{\theta \cdot x} \, \nu(dx) \le \int e^{-\sigma||x_{(1)}|| + \theta_{(2)} \cdot x_{(2)}} \, \nu(dx)$$

$$\downarrow \int e^{\theta_{(2)} \cdot x_{(2)}} \, \nu_{|F_0}(dx)$$

as $\sigma \to \infty$, uniformly for $\theta_{(2)} \le \delta_0$. In particular

$$(3) \qquad \sup \{|\psi(\theta)|: \ \theta \in Q(\sigma, \delta)\} \to \psi_{|F_0}(0) = 0$$

as $\sigma \to \infty$, $\delta \to 0$. It follows that

$$(4) \qquad \sup \ \{|p_\theta(x) - q_0(x)|: \ \theta \in Q(\sigma, \delta)\} \to 0 \quad \text{as} \quad \sigma \to \infty, \ \delta \to 0$$

for each $x \in K$. [For $x \in F_0$ the convergence in (4) is uniform over compact subsets of F_0; however if $x \notin F_0$ then as $\sigma \to \infty$, $\delta \to 0$, $p_\theta(x) = e^{\theta \cdot x - \psi(\theta)} \sim e^{\theta \cdot x} \to 0 = q_0(x)$, but the convergence is not uniform over arbitrary compact subsets of K. (It is uniform over bounded subsets of X if $e_1 \cdot x < -\varepsilon < 0$ for all $x \in X - F_0$.)]

It remains to show that for given $\sigma \geq \sigma_0$, $\delta \leq \delta_0$ there is an $\alpha > 0$ such that $||\xi|| < \alpha$, $\xi \in K^\circ$, implies $\theta(\xi) \in Q(\sigma, \delta)$. Once this has been done it follows from (4), and the induction hypothesis, that $q_\xi(x)$ is continuous in $\xi \in K$ for each $x \in K$.

For convenience we show below only that there is an $\alpha > 0$ such that $||\xi|| < \alpha$ implies $\theta(\xi) \in Q(0, \delta)$. The proof for arbitrary $\alpha > 0$, in place of $\sigma = 0$, requires only minor alterations of the constants appearing in the proof. In the following α, ε are generic positive constants whose numerical value may decrease as the proof progresses. Since $0 \in N_{|F_0}$ there is an $\alpha > 0$ such that $||\theta_{(2)}|| > \delta$ implies $\psi_{|F_0}(\theta_{(2)}) \geq 2\beta||\theta_{(2)}||$. Let $C \subset X$ be a finite subset of X such that $C \cap F_0 \neq \phi$ and $F \cap C \neq \phi$ for every face F of K which properly contains F_0. The existence of C is guaranteed by 6.22(1).

Suppose $||\theta_{(2)}|| > \delta$ and $\theta_{(1)} \cdot x_{(1)} > 0$ for some $x \in K$. Then $\max\,\{\theta_{(1)} \cdot x_{(1)}\colon\ x \in C\} > 0$. If $||\xi|| < \alpha$ and α is sufficiently small then $\xi_{(1)}$ is in the convex hull of $\{x_{(1)}\colon\ x \in C\} \cup \{0\}$. Hence there is an $\eta \in R$ such that

(5) $$\theta_{(1)} \cdot \xi_{(1)} \leq \eta\alpha \max\,\{\theta_{(1)} \cdot x_{(1)} \cdot x \in C\}$$

for all $||\xi|| < \alpha$. Let $\rho = \max\,\{||x_{(2)}||\colon\ x \in C\}$, $\nu_0 = \min\,\{\nu(\{x\})\colon\ x \in C\}$. Then

$$\ell(\theta, \xi) = \theta \cdot \xi - \psi(\theta)$$

$$= \theta_{(2)} \cdot \xi_{(2)} - \beta||\theta_{(2)}|| + \theta_{(1)} \cdot \xi_{(1)} - \ln(e^{-\beta||\theta_{(2)}||}\lambda(\theta)) .$$

Now,

(6) $$\lambda(\theta) \geq \lambda_{|F_0}(\theta_{(2)}) + \nu_0 \exp\,(\theta_{(1)} \cdot x_{(1)} + \theta_{(2)} \cdot x_{(2)})$$

$$\geq \exp\,(2\beta||\theta_{(2)}||) + \nu_0 \exp\,(\theta_{(1)} \cdot x_{(1)} - \rho||\theta_{(2)}||) .$$

For notational simplicity let $t = \theta_{(1)} \cdot x_{(1)} > 0$. Then for $\alpha \leq \beta/2$

$$(7) \quad \ell(\theta, \xi) \leq \theta_{(2)} \cdot \xi_{(2)} - \beta||\theta_{(2)}|| + \eta\alpha t - \ln (e^{\beta||\theta_{(2)}||} + \nu_0 \exp (t - \rho||\theta_{(2)}|| - \beta||\theta_{(2)}||)$$

$$\leq -\varepsilon + \eta\alpha t - (\beta||\theta_{(2)}|| \ \ V(t - (\rho + \beta)||\theta_{(2)}|| + \ln \nu_0))$$

$$\leq -\varepsilon$$

for $\alpha > 0$ sufficiently small, since

$$\beta||\theta_{(2)}|| \ V(t - (\rho + \beta)||\theta_{(2)}|| - a\delta) \geq \frac{\beta t}{\rho + 2\beta + a}$$

for $||\theta_{(2)}|| > \delta, \ a \geq 0$.

If $||\theta_{(2)}|| > \delta$ but $\theta_{(1)} \cdot x_{(1)} \leq 0$ for all $x \in K$ then $\theta_{(1)} \cdot \xi_{(1)} \leq 0$ and

$$(8) \quad \ell(\theta, \xi) \leq \theta_{(2)} \cdot \xi_{(2)} - \psi_{|F_0}(\theta_{(2)}) + \theta_{(1)} \cdot \xi_{(1)} - \ln\left(\frac{\lambda(\theta)}{\lambda_{|F_0}(\theta)}\right)$$

$$\leq \theta_{(2)} \cdot \xi_{(2)} - \psi_{F_0}(\theta_{(2)}) \leq -\varepsilon \ .$$

If $||\theta_{(2)}|| \leq \delta_1$ but $\theta_{(1)} \cdot x_{(1)} > 0$ for some $x \in K$ then $\theta_{(1)} \cdot x_{(1)} > 0$ for some $x \in C$; and

$$(9) \quad \ell(\theta, \xi) \leq \theta_{(2)} \cdot \xi_{(2)} - \psi_{|F_0}(\theta_{(2)}) + \eta\alpha\theta_{(1)} \cdot x_{(1)} - \ln\left(1 + \frac{\nu_0 e^{\theta_{(1)} \cdot x_{(1)}}}{\psi_{|F_0}(\theta_{(2)})}\right)$$

$$\leq -\varepsilon < 0$$

for $\alpha > 0$ and some $\varepsilon > 0$ sufficiently small, since $\psi_{|F_0}(\theta_{(2)}) \geq 0$ but $\sup \{\psi_{|F_0}(\theta_{(2)}) : \ ||\theta_{(2)}|| \leq \delta_1\} < \infty$. If $||\xi|| < \alpha$ and $\theta \notin Q$ one of (7), (8), or (9) apply so that

(10) $$\ell(\theta, \xi) \le -\varepsilon < 0 .$$

On the other hand, there is a $\sigma > 0$ sufficiently large so that by (2) or (3),

(11) $$\ell(\sigma e_1, \xi) = \sigma e_1 \cdot \xi - \psi(\sigma e_1) \ge \sigma e_1 \cdot \xi - \varepsilon/3$$

$$\ge -2\varepsilon/3$$

for $||\xi|| < \alpha \le \frac{\varepsilon}{3\sigma}$. It follows from (10) and (11) that if $||\xi|| < \alpha$, $\xi \in K°$, then if $\theta \notin Q$

$$\ell(\theta, \xi) \le -\varepsilon < -2\varepsilon/3 \le \ell(\theta(\xi), \xi).$$

Hence $\theta \ne \theta(\xi)$. It follows that $\theta(\xi) \in Q$.

We have thus proved that given σ, δ there is an $\alpha > 0$ such that $||\xi|| < \alpha$, $\xi \in K°$, implies $\theta(\xi) \in Q(\sigma, \delta)$. As previously noted, this completes the proof of the theorem. ||

EXERCISES

6.6.1

Assume ϕ is regularly strictly convex. Verify 6.6(3).

6.7.1

For ϕ regularly strictly convex, when does $d_\phi = \phi$?

6.9.1

Generalize Theorem 3.9 to apply to steep, regularly convex functions ϕ [i.e.; write $\phi = \begin{pmatrix}\phi_{(1)}\\ \phi_{(2)}\end{pmatrix}$ and consider the map $\theta \to \begin{pmatrix}\xi_{(1)}(\theta)\\ \phi_{(2)}(\theta)\end{pmatrix}$. Show this map is 1 - 1 and continuous on N° with range $\xi_{(1)}(N^\circ) \times \phi_{(2)}(N^\circ) = K_{(1)} \times \phi_{(2)}(N^\circ)$].

6.18.1

(i) Show that $K_{I.F} \neq F$ in the following example:

$X = (1, -1) \cup \{((i^2 - 1)^{\frac{1}{2}}/i,\ \ 1/i);\ \ i=1,2,\ldots\}$, $F = K \cap H((1, 0), 1)$.

(ii) Construct an example of the same phenomenon in R^3 where X is a discrete set (i.e. X has no accumulation points in R^3). [Construct X so that the set X in (i) is its projection on the space spanned by the first two coordinate axes.]

6.19.1

Show that the following three families are aggregate exponential families:

(i) Binomial (n, p), $0 \leq p \leq 1$

(ii) Poisson (λ), $\lambda \geq 0$

(iii) Multinomial $(N, \underset{\sim}{p})$, $0 \leq p_i$, $\sum_{i=1}^{k} p_i = 1$.

6.19.2

Suppose the distribution of $X^{(i)}$ form an aggregate exponential family $\{q_{\underset{\sim}{\xi}}^{(i)}\}$, i=1,2, and $X^{(1)}$, $X^{(2)}$ are independent. Show that the distributions of $(X^{(1)}, X^{(2)})$ form a ($k_1 + k_2$ parameter) aggregate exponential family.

6.20.1

Construct an example in which 6.18(3) holds but $F \neq$ conhull X. [Let X' be the set in 6.18.1(i) and define $X \in R^3$ by

$X = \{x\colon (x_1, x_2) \in X',\ \ x_3 = \pm(1 - x_2)\} \cup (1,0,1) \cup (1,0,-1).$]

6.21.1

Let X be the set defined in 6.20.1 with the additional point (1,0,0). Show

(i) 6.18(3) fails at $x = (1,0,0)$.

(ii) The maximum likelihood estimate for the aggregate family $\{q_\xi\colon\ \xi \in F\}$ fails to exist (i.e. is the empty set) when $X = (1,0,0)$, which occurs with positive probability.

(iii) The failure in (ii) can be rectified in a natural way by letting G = conhull $\{(1,0,-1),\ (1,0,1)\}$ and adding the densities $q_{\xi(\theta)|G} = p_{\theta|G}$ to the family $\{q_\xi\colon\ \xi \in F\}$.

(iv) Addition of the densities $q_{\xi|G}$ is "natural" in the sense that for each $\xi \in G$ there is a sequence $\theta_i \in N^\circ$ such that $q_{\xi|G}(x) = \lim_{i\to\infty} p_{\theta_i}(x)$. [This sequence cannot be chosen to be of the form $\theta_i = \theta' + iv$ for fixed $v \in R^k$, $\theta' \in N^\circ$ as was the case in the proof of Theorem 6.21.]

6.21.2

Let ν be linear measure on the perimeter ∂S, of the unit square, S. This measure does not have a countable supporting set. Nevertheless, describe its "natural aggregate family", having parameter space S and satisfying the conclusion of Theorem 6.21 for each $x \in S$.

6.21.3

(i) Let ν be uniform measure on the perimeter S, say, of the unit circle S. Thus, $\{p_\theta\}$ is the family of Von-Mises distributions (Example 3.8). Show there can be no possible way of constructing a family of densities $\{q_\xi\}$ which contains $\{p_\theta\}$ such that the maximum likelihood estimate for $\{q_\xi\}$ exists with probability one. [$\lim_{||\theta||\to\infty} p_\theta(x) = \infty$ for each $x \in \partial S$.]

(ii) Note that if $\bar{X}_n$ is the sample mean from a sample of size n, $n \geq 2$, having the above distribution, then the maximum likelihood estimate does exist with probability one.

(iii) Construct a measure ν for which $\{p_\theta\}$ is a regular exponential family but there does not exist an n for which it is possible to construct an "aggregate family" of densities $\{q_\xi\}$, containing the densities of $\bar{X}_n$ under θ, such that the maximum likelihood estimator exists with probability one. [There exists such a measure ν having $K_\nu = \{x \in R^3: \; x_2^2 + x_3^2 \leq x_1^2, \; 0 \leq x \; \leq 1\}$, and $\nu(\{0\}) > 0$.]

6.22.2

Show that 6.22(1) (including the polyhedral nature of K) implies 6.20(1). [The polyhedrality of K guarantees that for every $x \in \partial K$ there is a face F of K such that $x \in \text{ri } F$.]

6.22.2

Prove that 6.22(1) and 6.22(3) are satisfied whenever ν is a product measure on a countable set $X = \prod_{j=1}^{k} X_j, \; X_j \in R$. [The faces $F = H(v, \alpha) \cap X$ of X are determined uniquely by $(\text{sgn } v_1,\ldots,\text{sgn } v_k)$.]

6.22.3

(i) Prove that

(1) $$N_{|F} = \text{Proj}_{\text{csp } F}(N_{|F}) \times (\text{csp } F)^\perp \; , \quad \text{and}$$

(2) $$\text{Proj}_{\text{csp } F}(N) \subset \text{Pr}_{\text{csp } F}(N_{|F}) \; .$$

(ii) Give an example in which $X = \{0,1,\ldots\}^2$,

$$F = \{(0, 0), (1, 0),\ldots\}, \qquad N = (-\infty, 0)^2 \; ,$$

(3) $$\text{Proj}_{\text{csp } F}(N) = (-\infty, 0) \times 0 \neq R \times 0 = \text{Proj}_{\text{csp } F}(N_{|F}) \; ,$$

and

(4) $$\xi_{|F}((0,\ 0)) = (1,\ 0) \in X\ .$$

(Thus 6.22(3) is not valid here.)

(ii) In the example (ii) show that $q_\xi((x_1, 0))$, $x_1 = 0, 1, \ldots$, is not continuous at $\xi = (\xi_1, 0)$, $\xi_1 > 1$. [If θ_i is chosen so that $\theta_{i1} \uparrow 0$ somewhat slowly and $\theta_{i2} \to -\infty$ then $\xi(\theta_i) \to (\xi_1, 0)$ but $q_{\xi(\theta_i)}(x) \nrightarrow q_{(1,0)}(x)$.]

<u>6.23.1</u>

Prove versions of Theorems 5.7, 5.8 and 5.12 valid for aggregate exponential families. [Make the assumptions in Theorem 6.23.]

<u>6.23.2</u>

Show that $q_\xi(x)$ is not jointly continuous in (ξ, x) at any point with $\xi = x \in \partial K$.

<u>6.23.3</u>

Are the analogs to Theorems 6.12 and 6.13 valid for aggregate exponential families under the assumptions of Theorem 6.23?

<u>6.23.4</u>

Suppose $X = (0, 0) \cup \{x \in R^2:\ x_i = 1, \ldots,\ i = 1,2\}$. Note that Assumption 6.22(1) is not satisfied. Show that, nonetheless, $q_\xi(x)$ is continuous at every $\xi \in$ conhull $X = F$. (If one *defines* $q_\xi(x) = q_0(x)$ for $\xi \in K$ - conhull X then it is even true that $q_{\cdot}(x)$ is continuous on K.)

<u>6.23.5</u>

Let $X = \{((i^2 - 1)^{\frac{1}{2}}/i,\ 1/i):\ i = 1, \ldots\} \cup (1,\ 0)$. For $x = ((i^2 - 1)^{\frac{1}{2}}/i,\ 1/i) \in X$ let $\nu(\{x\}) = 1/2^i$, and let $\nu(\{0\}) = 1$. Note that 6.22(1) is not satisfied. Show that $q_\xi((1,0))$ is not continuous at $\xi = (1,0)$ [$q_{(1,0)}((1,0)) = 1$. Let $0 < c < 1$. For ℓ sufficiently large let $\theta_\ell = \rho_\ell x_\ell$ with ρ_ℓ chosen so that $p_{\theta_\ell}((1,0)) = c$ ($\{\rho_\ell\}$ is a swiftly increasing sequence.) Then $\xi(\theta_\ell) \to (1, 0)$ but $q_{\xi(\theta_\ell)}((1, 0)) \equiv c \neq 1$.] (In this example $q_\xi(0)$ is, however, upper semicontinuous; so that, for example, the conclusion of Theorem 6.23 remains valid. Exercise 6.23.4 shows this need not be the case.)

6.23.6

For $x = x^{(ij)} = ((i^2 - 1)^{\frac{1}{2}}/i\ ,\ 1/i,\ j),\ i=1,\ldots,\ j = \pm 1$, let $\nu(\{x^{(ij)}\}) = (4 + 3j)/2^i$. For $x = x^{(j)} = (1, 0, j)$, $j=-1, 0, +1$ let $\nu(\{x\}) = 2 - |j|$. Otherwise $\nu(\{x\}) = 0$.

Construct $\{\theta_\ell\}$ in a manner similar to 6.23.5 with $(\theta_\ell)_3 = 0$ so that $P_{\theta_\ell}(\{x^{(j)}\colon\ j=0, \pm 1\}) \uparrow 1/3$ and $(\xi(\theta_\ell))_1 \to 1$. Verify that $\xi(\theta_\ell) \to (1, 0, 1/2)$ and $P_{\theta_\ell}(\{x^{(-1)}\}) = p_{\theta_\ell}(x^{(-1)}) \uparrow 1/12$, but $q_{(1,0,\frac{1}{2})}(x^{(-1)}) = P_{(1,0,\frac{1}{2})}(x^{(-1)}) = (1/4)^2 < 1/12$. Hence $q_\xi(x^{(-1)})$ is not continuous at $\xi = (1, 0, 1/2)$ or even upper semicontinuous. If $E \subset K$ is the closed set $\{\xi(\theta_\ell)\colon \ell=1,\ldots\} \cup (1, 0, 1/2)$ then the maximum likelihood estimator over the family $\{q_\xi\colon \xi \in E\}$ fails to exist at the possible observation $x^{(-1)}$.

CHAPTER 7. TAIL PROBABILITIES

In exponential families the probability under θ of a set generally falls off exponentially fast as the distance of the set from $\xi(\theta)$ increases. This section contains several results of this form. The first of these will be improved later, but it is included here because of its simplicity of statement and proof.

Throughout this chapter let $\{p_\theta\}$ be a steep canonical exponential family. (Most of the results hold with possibly minor modifications for non-minimal families, and many also hold for non-steep families.)

FIXED PARAMETER (Via Chebyshev's Inequality)

7.1 Theorem

Fix $\theta_0 \in N^\circ$. Choose ε so that $\{\theta: ||\theta - \theta_0|| \leq \varepsilon\} \subset N^\circ$. Then there exists a constant $c < \infty$, such that

(1) $$\Pr_{\theta_0} H^+(v, \alpha) < c \exp(-\varepsilon\alpha)$$

for all $v \in R^k$ with $||v|| = 1$ and all $\alpha \in R$.

Proof. Let

(2) $$c = \exp(\sup \{\psi(\theta) - \psi(\theta_0): ||\theta - \theta_0|| = \varepsilon\})$$

and let $\theta_\varepsilon = \theta_0 + \varepsilon v$. Then

$$\Pr_{\theta_0}\{H^+(v,\alpha)\} = \int_{H^+(v,\alpha)} \exp(\theta_0 \cdot x - \psi(\theta_0))\nu(dx)$$

$$= \int_{H^+(v,\alpha)} \exp(\theta_0 \cdot x + (\varepsilon v) \cdot x - (\varepsilon v) \cdot x - \psi(\theta_0))\nu(dx)$$

$$\leq \left(\int_{H^+(v,\alpha)} \exp(\theta_\varepsilon \cdot x - \psi(\theta_\varepsilon))\nu(dx)\right)\exp(\psi(\theta_\varepsilon) - \psi(\theta_0) - \varepsilon\alpha)$$

$$\leq c \exp(-\varepsilon\alpha) \quad . \qquad ||$$

Note that (2) provides a specific formula for the constant appearing in (1).

In specific situations the bound provided in Theorem 7.1 can be improved in various ways. However the following converse result shows that Theorem 7.1 always comes within an arbitrarily small amount of yielding the best exponential rate of decrease for tail probabilities.

7.2 Proposition

Let $\theta_0 \in N^\circ$. Suppose there exists a $c < \infty$ and $\varepsilon > 0$ such that 7.1(1) is valid for all $v \in R^k$ with $||v|| = 1$ and all $\alpha > 0$. Then $\{\theta: ||\theta - \theta_0|| < \varepsilon\} \subset N^\circ$.

(Thus, if for some $\varepsilon > 0$, $c < \infty$, a bound of the form 7.1(1) is valid for all v with $||v|| = 1$ and all $\alpha > 0$, then Theorem 7.1 will verify such a bound for any $\varepsilon' < \varepsilon$.)

Proof. We leave the proof as an exercise. ||

When $\varepsilon = \inf \{||\theta - \theta_0|| : \theta \notin N\}$ then 7.1(1) may or may not be valid for all α, ν. The following example demonstrates this.

7.3 Example

Relative to Lebesgue measure, let

$$(1)\qquad f_{\sigma,k}(y) = \begin{cases} \Gamma(k)y^{k-1}\, e^{-y/\eta}/\eta^k & y > 0 \\ 0 & y \le 0 \end{cases} .$$

This is the gamma density with scale parameter η and shape parameter k. Let $x_1 = y$, $x_2 = \ln y$, $\theta_1 = -1/\eta$, $\theta_2 = (k - 1)$, and let ν be the measure induced by the map $y \to x$ when y has Lebesgue measure on $(0, \infty)$. One then has a standard exponential family of order 2 with

$$\psi(\theta) = (\theta_2 + 1)\ln(-\theta_1) - \ln\Gamma(\theta_2 + 1)$$

and

$$(2)\qquad N = (-\infty, 0) \times (-1, \infty), \qquad K = \{(x_1, x_2):\ x_1 \ge 0,\ \ x_2 \ge \ln x_1\} .$$

When $k = 1$ (i.e. $\theta_2 = 0$) the resulting one-parameter exponential family is that of exponential distributions with intensity $|\theta_1|$. For this family

$$\Pr_{\theta_1=-1}\{x_1 > \alpha\} = e^{-\alpha} \qquad \text{for all} \qquad \alpha > 0$$

so that 7.1 holds with $v = 1$ and $\varepsilon = 1 = \inf\{||\theta - \theta_0|| : \theta \notin N\}$. On the other hand, for $\theta_2 = 1$ the resulting one-parameter gamma family has

$$\Pr_{\theta_1=-1}\{x_1 > \alpha\} = (\alpha + 1)e^{-\alpha} \qquad \text{for all} \qquad \alpha > 0.$$

Thus here 7.1(1) fails to hold when $v = 1$ and $\varepsilon = 1 = \inf\{||\theta - \theta_0|| : \theta \notin N\}$

When $N = R^k$ Theorem 7.1 says only that $\Pr_{\theta_0}\{H^+(u, \alpha)\} = O(e^{-k\alpha})$ for all $k > 0$. However, much smaller bounds may be valid for these tail probabilities. Consider for example the following well known facts:

$$(3)\qquad \int_\alpha^\infty e^{-t^2/2}\, dt \le e^{-\alpha^2/2}/\alpha \qquad \text{for} \qquad \alpha > 0$$

and

(4) $$\int_{\alpha}^{\infty} e^{-t^2/2}\,dt \sim e^{-\alpha^2/2}/\alpha \quad \text{as} \quad \alpha \to \infty \quad .$$

Thus, suppose X is normal, mean 0, variance 1. Then, from (3)

(5) $$\Pr\{X > \alpha\} \le e^{-\alpha^2/2}/\alpha(2\pi)^{\frac{1}{2}} \quad \text{for} \quad \alpha > 0 \quad .$$

It can be seen from (4) that this bound is asymptotically accurate as $\alpha \to \infty$.

Theorem 7.5 contains a bound which easily yields the statement

(6) $$\Pr\{X > \alpha\} \le e^{-\alpha^2/2}$$

for this situation. This is much better than what is available from 7.1(1) but is still inferior to (5).

Theorem 7.1 applies to probabilities of large deviations defined by half spaces but can easily be converted to a statement about any shape of set, as follows.

7.4 Corollary

Consider a standard exponential family. Fix $\theta_0 \in N^{\circ}$. Let $x_0 \in R^k$. Let S be any set. Let $\rho = \inf\{||x - x_0|| : x \notin S\}$, and define ε as in Theorem 7.1. Then there is a $c < \infty$ such that

(1) $$P_{\theta_0}(\{(X - x_0)/\alpha \notin S\}) < c\exp(-\varepsilon\rho\alpha) \quad \text{for all} \quad \alpha \in R \quad .$$

Proof. It suffices to prove the corollary for $x_0 = 0$ and S the open sphere of radius ρ about the origin.

There exists $\rho' < \rho$ and $\varepsilon' < \inf\{||\theta - \theta_0|| : \theta \notin N\}$ such that $\varepsilon'\rho' = \varepsilon\rho$. There exists a finite set of unit vectors $\{a_i: i=1,\ldots,n\}$ such that $\bigcap_{i=1}^{n} \{x: x \cdot a_i < \rho'\} \subset S$. Thus $\Pr_{\theta_0}\{X/\alpha \notin S\} < \sum_{i=1}^{n} \Pr_{\theta_0}\{X \cdot a_i > \alpha\rho'\}$

$\le \sum_{i=1}^{n} c_i \exp(-\alpha\rho'\varepsilon') \le c\exp(-\varepsilon\rho\alpha)$ by Theorem 7.1 where $c < \infty$ is an appropriate constant. ||

FIXED PARAMETER (Via Kullback-Leibler Information)

It is possible to use the Kullback-Leibler information number (i.e. entropy) to improve the preceding bound. See the exercises for some applications of this bound to asymptotic theory.

7.5 Theorem

Let $\theta_0 \in N^\circ$ and $\bar{H}^+ = \bar{H}^+(v, \alpha)$. Then

(1) $$P_{\theta_0}(\bar{H}^+) \leq \exp(-\tilde{K}(\bar{H}^+, \xi(\theta_0))) \quad .$$

Proof. Suppose first that

(2) $$\bar{H}^+ \cap K^\circ \neq \phi \ .$$

Let $\tilde{\xi} = \tilde{\xi}_{\bar{H}^+}(\theta_0)$. Note that $\tilde{\xi} \in \bar{H}^+ \cap K^\circ$ by Theorem 6.13. Hence $\tilde{\theta} = \theta(\tilde{\xi}) \in N^\circ$. (This is precisely the situation pictured in Figure 6.14(1).) Now,

(3) $$\tilde{k} = \tilde{K}(\bar{H}^+, \xi(\theta_0)) = (\tilde{\theta} - \theta_0) \cdot \tilde{\xi} - \psi(\tilde{\theta}) + \psi(\theta_0)$$

$$\leq (\tilde{\theta} - \theta_0) \cdot x - \psi(\tilde{\theta}) + \psi(\theta_0) \qquad \forall \ x \in \bar{H}^+$$

by definition and by 6.13(2). This yields

$$P_{\theta_0}(\bar{H}^+) = \int_{\bar{H}^+} p_{\theta_0}(x)\nu(dx) = \int_{\bar{H}^+} \frac{p_{\theta_0}(x)}{p_{\tilde{\theta}}(x)} p_{\tilde{\theta}}(x)\nu(dx)$$

$$= \int_{\bar{H}^+} \exp((\theta_0 - \tilde{\theta}) \cdot x - \psi(\theta_0) + \psi(\tilde{\theta}))p_{\tilde{\theta}}(x)\nu(dx)$$

$$\leq \int_{\bar{H}^+} \exp(-\tilde{k})p_{\tilde{\theta}}(x)\nu(dx) \leq e^{-\tilde{k}} \quad ,$$

which is the desired result.

Now suppose $\bar{H}^+ \cap K \neq \phi$ but $\bar{H}^+ \cap K^\circ = \phi$. Then

(4) $$\lim_{\varepsilon\downarrow 0} \tilde{K}(\bar{H}^+(v, \alpha - \varepsilon), \xi(\theta_0)) = \tilde{K}(\bar{H}^+(v,\alpha), \xi(\theta_0)) \le \infty$$

since $\tilde{K}(\cdot, \xi(\theta_0))$ is lower semi-continuous (by definition), satisfies

$$\lim_{||\xi||\to\infty} \tilde{K}(\xi, \xi(\theta_0)) = \infty$$

(by 6.5(5)), and since $\tilde{K}(\bar{H}^+(v, \alpha), \xi(\theta_0)) \ge \tilde{K}(\bar{H}^+(v, \alpha - \varepsilon), \xi(\theta_0))$ for all $\varepsilon > 0$. Hence

(5) $$P_{\theta_0}(\bar{H}^+) = \lim_{\varepsilon\downarrow 0} P_{\theta_0}(\bar{H}^+(v, \alpha-\varepsilon)) \le \lim_{\varepsilon\downarrow 0} \exp(-\tilde{K}(\bar{H}^+(v, \alpha-\varepsilon), \xi(\theta_0)))$$

$$= \exp(-\tilde{K}(\bar{H}^+, \xi(\theta_0))) \quad . \qquad ||$$

(We leave as an exercise to verify that

(6) $$\tilde{K}(\bar{H}^+, \xi(\theta_0)) = \infty \quad \text{if and only if} \quad P_{\theta_0}(\bar{H}^+) = 0 \quad .)$$

Note that the Kullback-Leibler information enters into the above only as a convenient way of identifying the sup $\{(\tilde{\theta} - \theta_0) \cdot x - \psi(\tilde{\theta}) + \psi(\theta_0)$: $x \in H^+\}$. Various other interpretations of K, such as the probabilistic Definition 6.1, do not enter into the above argument.

The connection between Theorem 7.5 and 7.1 is provided by the following lemma.

<u>7.6 Lemma</u>

Let $\theta_0 \in N°$ and $H^+ = H^+(v, \alpha)$. Suppose $\theta = \theta_0 + \varepsilon v \in N°$. Then

(1) $$\tilde{K}(H^+, \xi(\theta_0)) \ge \psi(\theta_0) - \psi(\theta) + \varepsilon\alpha \quad .$$

Proof. Let $\tilde{\xi} = \tilde{\xi}_{\bar{H}^+}(\theta_0)$ as in Theorem 7.5. Then

$$\tilde{K}(H^+, \xi(\theta_0)) = (\tilde{\theta} - \theta_0) \cdot \tilde{\xi} + \psi(\theta_0) - \psi(\tilde{\theta})$$

$$\ge (\theta - \theta_0) \cdot \tilde{\xi} + \psi(\theta_0) - \psi(\theta)$$

since $\tilde{\theta} = \theta(\tilde{\xi}) = \hat{\theta}_N(\tilde{\xi})$ maximizes $\ell(\cdot, \tilde{\xi})$. Hence

$$\tilde{K}(H^+, \theta_0) \geq \varepsilon v \cdot \tilde{\xi} + \psi(\theta_0) - \psi(\theta) = \varepsilon\alpha + \psi(\theta_0) - \psi(\theta) . \quad ||$$

Applying the bound (1) in the formula 7.5(1) yields the earlier formulae, 7.1(1) and (2), of Theorem 7.1.

Note also that in the normal example of Example 7.3, $\tilde{K}(\xi, 0) = \xi^2/2$, and thus 7.5(1) yields 7.3(6).

FIXED REFERENCE SET

The preceding results concern the nature of probabilities of large deviations when the parameter is fixed and the reference set for calculating the probability proceeds to infinity. There is another class of results. These concern the situation when the reference set is fixed and the parameter proceeds to infinity in an appropriate direction. These theorems were exploited in a statistical setting by Birnbaum (1955) and then Stein (1956). Giri (1977) surveys several further applications of this theory.

7.7 Theorem

Let $v \in R^k$, $\alpha \in R$. Let $S_1, S_2 \subset R^k$ with

(1) $$S_2 \subset \bar{H}^-(v, \alpha) ,$$

(2) $$\nu(S_1 \cap H^+(v, \alpha)) > 0 .$$

Let $K \subset N$ be compact. Then there exist constants c and $\varepsilon > 0$ such that

(3) $$\frac{\int_{S_2} e^{\theta \cdot x}\,\nu(dx)}{\int_{S_1} e^{\theta \cdot x}\,\nu(dx)} \leq c \exp(-\rho\varepsilon)$$

for all $\theta \in N$ of the form $\theta = \eta + \rho v$ with $\eta \in K$, $\rho > 0$.

Proof. Let $S_1(\varepsilon) = S_1 \cap H^+(v, \alpha + \varepsilon)$. There is an $\varepsilon > 0$ such that $\nu(S_1(\varepsilon)) > \varepsilon > 0$. Then,

$$\frac{\int_{S_2} e^{\theta \cdot x}\, \nu(dx)}{\int_{S_1} e^{\theta \cdot x}\, \nu(dx)} \le \frac{\int_{S_2} \exp(\rho(v \cdot x - \alpha) + \rho\alpha + \eta \cdot x)\nu(dx)}{\int_{S_1(\varepsilon)} \exp(\rho(v \cdot x - \alpha) + \rho\alpha + \eta \cdot x)\nu(dx)}$$

$$\le \frac{\int_{S_2} e^{\eta \cdot x}\, \nu(dx)}{e^{\rho\varepsilon} \int_{S_1(\varepsilon)} e^{\eta \cdot x}\, \nu(dx)} \le c \exp(-\rho\varepsilon)$$

where

$$c = \sup_{\eta \in K} \left(\int_{S_2} e^{\eta \cdot x}\, \nu(dx) \Big/ \int_{S_1(\varepsilon)} e^{\eta \cdot x}\, \nu(dx)\right) < \infty \quad . \tag{4}$$

Here is why $c < \infty$: K is compact and $\nu(S_1(\varepsilon)) > 0$ so that $\inf_{\eta \in K} \int_{S_1(\varepsilon)} e^{\eta \cdot x}\, \nu(dx) > 0$. Also, $\int_{S_2} e^{\eta \cdot x}\, \nu(dx)$ is upper semicontinuous on K by Fatou's lemma, and is finite on K since $K \subset N$. Thus $\sup_{\eta \in K} \int_{S_2} e^{\eta \cdot x} \nu(dx) < \infty$. ||

The preceding theorem really concerns the relationship of probabilities for the sets S_2 and $S_1(0) = S_1 \cap H^+(v, \alpha)$ contained in separate half spaces. Note again the dual relationship, connecting $\theta \in N$ and $H \subset K$ in Theorem 7.7. Because of this relationship it is often revealing in such contexts to superimpose both the sample space and parameter space on a single plot. This is done in Example 7.12(1).

Here are some corollaries to the Theorem, the second of which will be used in the example. The first of these corollaries may be instructively compared to Theorem 7.1.

7.8 Corollary

Let $v \in R^k$, $K \subset N$ be compact, and $S \subset \bar{H}^-(a, \alpha)$. Suppose

(1) $$\nu(H^+(v, \alpha)) > 0 \quad .$$

Then there exist constants c and $\varepsilon > 0$ such that

$$\Pr_\theta(S) \le c \exp(-\rho\varepsilon)$$

for all $\theta \in N$ of the form $\theta = \eta + \rho v$ with $\eta \in K$, $\rho > 0$. In particular, for any sequence $\{\theta_i \in N: \theta_i = \rho_i v + \eta_i, \; \rho_i \to \infty, \; \eta_i \in K\}$ one has

$$\lim_{i\to\infty} \Pr_{\theta_i}(S) = 0 \quad .$$

Proof. Let $S_2 = H^+(v, \alpha)$. Then by Theorem 7.7

$$\Pr_\theta(S) \le c \exp(-\rho\varepsilon) \int_{S_2} e^{\theta\cdot x-\psi(\theta)} \nu(dx)$$

$$= c \exp(-\rho\varepsilon)\Pr_\theta(S_2) \le c \exp(-\rho\varepsilon) \quad . \quad ||$$

7.9 Corollary

Again, let $v \in R^k$, $K \subset N°$ be compact, and $\nu(S) > 0$; and let $\{\theta_i\}$ be any sequence of the form $\theta_i = \rho_i v + \eta_i$ with $\rho_i \to \infty$ and $\eta_i \in K$. Then

(1) $$\lim_{i\to\infty} E_{\theta_i}(v \cdot X) = \sup\{\alpha: \; \nu(H^+(v, \alpha)) > 0\} \le \infty \quad .$$

(Note that here we assume $K \subset N°$; not merely $K \subset N$.)

Proof. Let α_0 denote the supremum on the right of (1). Since $E_{\theta_i}(v\cdot X) \le \alpha_0$ it is only necessary to prove $\liminf_{i\to\infty} E_{\theta_i}(v\cdot X) \ge \alpha_0$. To this end, let $\alpha < \alpha' < \alpha_0$ and $S_2 = H^-(v, \alpha')$. Let $\xi_2(\theta) = E_\theta(X|X \in S_2)$. If $\nu(S_2) = 0$ the result is trivial. Hence, suppose $\nu(S_2) > 0$. Note that $\xi_2(\theta)$ exists and is continuous for all $\theta \in N°$. Hence $\beta = \inf\{v \cdot \xi_2(\eta): \eta \in K\} > -\infty$. Note that

$\beta \leq \alpha'$.

Apply Corollary 2.5 to the conditional exponential family given $X \in S_2$ (generated by $\nu|_{S_2}$) to find

$$E_\theta(v \cdot X | X \in S_2) \geq E_\eta(v \cdot X | X \in S_2) \geq \beta$$

for all $\theta = \eta + \rho v$ with $\rho \geq 0$. Then for such θ,

$$E_\theta(v \cdot X) = Pr_\theta(X \in S_2) \cdot E_\theta(v \cdot X | X \in S_2)$$

$$+ Pr_\theta(X \notin S_2) \cdot E_\theta(v \cdot X | X \in S - S_2)$$

$$\geq (ce^{-\varepsilon\rho}/(1 + ce^{-\varepsilon\rho}))\beta + (1/(1 + ce^{-\varepsilon\rho}))\alpha'$$

by Theorem 7.7. Hence $E(v \cdot X) > \alpha$ (since $\alpha < \alpha'$) for θ as above for all ρ sufficiently large. This implies $\liminf_{i\to\infty} E_{\theta_i}(v \cdot X) \geq \alpha_0$, since $\alpha < \alpha_0$ was arbitrary. ||

Note the placement of the hyperplane H in the statement of Theorem 7.7. If $S_2 \subset H^-$ and $\nu(S_1 \cap \bar{H}^+) > 0$, but $\nu(S_1 \cap H^+) = 0$, then only a much weaker conclusion is valid. This conclusion is contained in the following corollary.

7.10 Corollary

Let $v \in R^k$, $\alpha \in R$. Suppose

(1) $$S_2 \subset H^-(v, \alpha)$$

and

(2) $$\nu(S_1 \cap \bar{H}^+(v, \alpha)) > 0 .$$

Let $K \subset N$ be compact. Let $\{\theta_i\} \subset N$ be a sequence of the form $\theta_i = \rho_i v + \eta_i$ with $\eta_i \in K$, $\rho_i \to \infty$. Then

(3) $$\lim_{i\to\infty} \frac{P_{\theta_i}(S_2)}{P_{\theta_i}(S_1)} = 0 .$$

Proof. Apply Theorem 7.7 to find

$$\lim_{i\to\infty} \frac{P_{\theta_i}(S_2 \cap H^-(v, \alpha - \varepsilon))}{P_{\theta_i}(S_1)} = 0 \tag{4}$$

for all $\varepsilon > 0$. Furthermore, if $\rho_i > 0$

$$\frac{P_{\theta_i}(S_2 \cap \bar{H}^+(v, \alpha - \varepsilon))}{P_{\theta_i}(S_1)} < \frac{P_{\eta_i}(S_2 \cap \bar{H}^+(v, \alpha - \varepsilon))}{P_{\eta_i}(S_1)} \to 0 \tag{5}$$

as $\varepsilon \to 0$ uniformly for $\eta_i \in K$.

(The inequality in (5) follows after applying Corollary 2.23 to the functions $h_c(x) = \chi_{S_1} - c\chi_{S_2 \cap \bar{H}^+(v,\alpha-\varepsilon)}$ with c chosen so that $E_{\eta_i}(h_c(X)) = 0$ to find that $E_{\theta_i}(h_c(X)) \geq E_{\eta_i}(h_c(X))$ for all c and $\rho_i > 0$.) Combining (4) and (5) yields the conclusion of the corollary. ||

7.11 Example

Consider the usual sufficient statistics $\bar{X}$, S^2 derived from a normal (μ, σ^2) sample. As explained in Example 1.2 the statistics $X_1 = \bar{X}$, $X_2 = S^2 + \bar{X}^2$ are the canonical statistics for a two-parameter exponential family with canonical parameters $\theta_1 = n\mu/\sigma^2$, $\theta_2 = -n/2\sigma^2$. Note that $K = \{(x_1, x_2): x_2 \geq x_1^2\}$. For some $c > 1$ consider the conditioning set $Q = \{(x_1, x_2): x_2 \geq cx_1^2\} = \{(\bar{x}, s^2): \bar{x}^2/s^2 \leq 1/(c - 1)\}$. (This is the set on which the usual two-sided t-test (based on $t = \sqrt{n-1}\,\bar{x}/s$) with n - 1 degrees of freedom accepts at the appropriate level determined by c.) Fix $\mu = \mu_0$ and let $\sigma^2 \to 0$. Then $(\theta_1, \theta_2) = (n/\sigma^2)(\mu_0, -\frac{1}{2})$. Thus (θ_1, θ_2) proceeds down the ray with slope $-\frac{1}{2}\mu_0$ as $\sigma^2 \to 0$. Both X and Θ are displayed on the plot in Figure 7.11(1), which shows also K, Q, and this line.

Corollary 7.9 applied to the conditional exponential family given $X \in Q$ (generated by the measure ν restricted to Q) yields

(1) $$\lim_{\sigma^2\to 0} E_{(\mu_0,\sigma^2)}(\mu_0 X_1 - X_2/2 | X \in Q)$$

$$= \sup \{\mu_0 x_1 - x_2/2 | (x_1, x_2) \in Q\} = \mu_0/2c^2 .$$

Note that $E(\mu_0 X_1 - X_2/2 | X \in Q) = (\mu_0, -\tfrac{1}{2}) \cdot E((X_1, X_2)|X \in Q)$ and that $E((X_1, X_2)|X \in Q) \in Q$. Furthermore since Q is strictly convex $(\mu_0, -\tfrac{1}{2}) \cdot (x_{1i}, x_{2i}) \to \mu_0/2c^2 = \sup \{(\mu_0, -\tfrac{1}{2}) \cdot (x_1, x_2): x_1, x_2 \in Q\}$ for a sequence $\{(x_{1i}, x_{2i})\} \subset Q$ if and only if $(x_{1i}, x_{2i}) \to (\mu_0/c, \mu_0^2/c)$. (Note that the tangent to Q at the point $(\mu_0/c, \mu_0^2/c)$ is perpendicular to the ray $(n/\sigma^2)(\mu_0, -\tfrac{1}{2})$.) Thus

(2) $$\lim_{\sigma^2\to 0} E_{(\mu_0,\sigma^2)}((X_1, X_2)|X \in Q) = (\mu_0/c, \mu_0^2/c) = e_0 \quad \text{(say)} .$$

In terms of the traditional variables $\bar{X}$, S^2, and $t = \sqrt{n-1}\,\bar{x}/s$ this yields

(3) $$\lim_{\sigma^2\to 0} E_{(\mu_0,\sigma^2)}((\bar{X}, S^2)|\ |t| \le \tau) = \left(\frac{\tau\mu_0}{\tau + n - 1}, \frac{(n-1)\mu_0^2}{(\tau + n - 1)^2}\right)$$

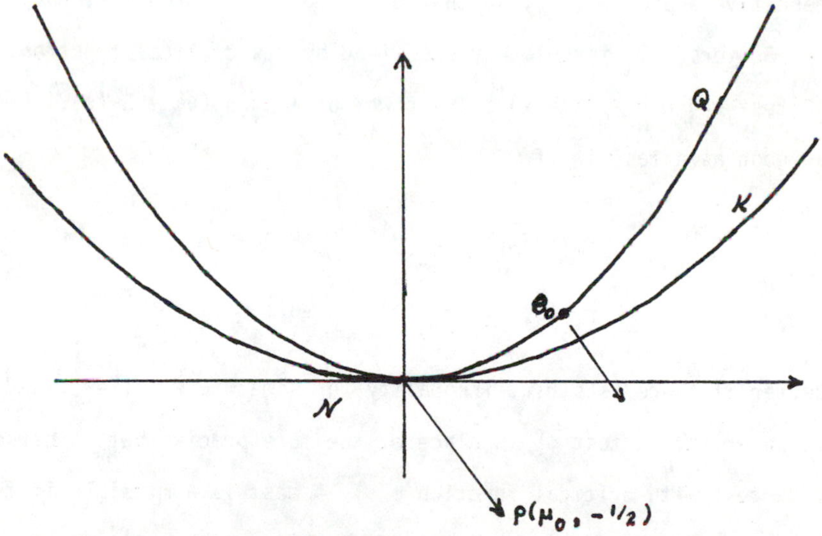

Example 7.11(1): Picture for Example 7.12

COMPLETE CLASS THEOREMS FOR TESTS (Separated Hypotheses)

The preceding results can be used to prove admissibility of many conventional test procedures in univariate and multivariate analysis of variance and in many other testing situations involving exponential families. When combined with the continuity theory for Laplace transforms of Section 2.17 these results yield useful complete class characterizations for certain classes of problems. In many of these cases the characterization precisely describes the minimal complete class. The general theory, as well as a very few specific applications, is described in the remainder of this chapter. Many more applications can be found in the cited references. The results to follow should be compared to the results in the same spirit for estimation which appear in Chapter 4.

7.12 Setting and Definitions

Throughout the remainder of this chapter $\{p_\theta: \theta \in \Theta\}$ is a standard exponential family. The parameter space Θ is divided into non-empty null and alternative spaces Θ_0, Θ_1; so that $\Theta = \Theta_0 \cup \Theta_1$. In the customary fashion, a test of Θ_0 versus Θ_1 is uniquely specified by its critical function, ϕ, where $\phi(x) = P(\text{test rejects } \Theta_0 | X = x)$. The power of ϕ is $\pi_\phi(\theta) = E_\theta(\phi)$. A test ϕ_1 is as good as a test ϕ_2 if

$$(1) \qquad \begin{aligned} \pi_{\phi_1}(\theta) &\le \pi_{\phi_2}(\theta) \qquad \theta \in \Theta \\ \pi_{\phi_1}(\theta) &\ge \pi_{\phi_2}(\theta) \qquad \theta \in \Theta \end{aligned}$$

It is better if there is strict inequality for some $\theta \in \Theta$. (Here, and in what follows, we write, "a test ϕ" in place of the more precise but cumbersome phrase, "a test with critical function ϕ".) A test is admissible if there is no better test. The decision-theoretic formulation with a two-point action space $A = \{a_0, a_1\}$ and a loss function of the form

$$L(\theta, a_j) = \ell(\theta) > 0 \text{ if } \theta \notin \Theta_j, \ = 0 \text{ if } \theta \in \Theta_j ,$$

yields the same ordering among tests, and hence the same collection of admissible tests.

Let

(2) $$U_r = U_r(\Theta, \theta_0)$$

$$= \left\{u:\ ||u|| = 1,\ \ \exists\ \theta \in \Theta \ni ||\theta|| > r,\ \text{and}\ u = \frac{\theta - \theta_0}{||\theta - \theta_0||}\right\},$$

$$r \geq 0\ ;$$

and let

(3) $$U(\Theta, \theta_0) = \bigcap_{r\geq 0} U_r(\Theta, \theta_0) \quad \text{and} \quad U^*(\Theta, \theta_0) = \bigcap_{r\geq 0} \bar{U}_r(\Theta, \theta_0) \quad .$$

Note that if Θ is a closed cone then $U = U^*$; more generally $U \subset U^*$. It is possible that $U = \phi$ but $U^* \neq \phi$.

If $S \subset R^k$ is a convex set let

(4) $$\alpha(u) = \alpha_S(u) = \sup\{x \cdot u:\ x \in S\} \quad .$$

This function is defined for $u \in R^k$,, although we will mainly be interested in its values for $||u|| = 1$. As is well known,

(5) $$\bar{S} = \bigcap_{\{u:||u||=1\}} \bar{H}^-(u, \alpha_S(u)) \quad .$$

It is clear from the definition (4) that $\alpha(\cdot)$ is lower semicontinuous.

The following lemma is a key result which leads directly to the first main theorem. A result of this type was first proved and used by Birnbaum (1955) in the case of testing for a normal mean. A general result similar to the following lemma was then proved and applied in Stein (1956b).

7.13 Lemma

Fix $\theta_2 \in R^k$. Let

(1) $$S = \bigcap_{u \in U^*} \bar{H}^-(u, \alpha_S(u))$$

where $U^* = U^*(\Theta_1, \theta_2)$. Assume further either that

(2) $$S = \bigcap_{u \in U} \bar{H}^-(u, \alpha_S(u)) , \qquad (U = U(\Theta, \theta_2)) ,$$

or $\alpha_S(u)$ is continuous at u for all $u \in U^* - U$. Let $\phi_1(x) = 1$ for all $x \notin S$. Suppose ϕ_2 is as good as ϕ_1. Then $\phi_2(x) = 1$ for $x \notin S$, a.e. (ν).

(Note: A more formal way to state the conclusion of the lemma is $\nu\{x: x \notin S, \phi_2(x) < 1\} = 0$.)

Proof. Assume for convenience $\theta_2 = 0$. Suppose the conclusion of the lemma is false. Then there is an $\varepsilon_0 > 0$, $u_0 \in U^*$ such that

(3) $$C_0 = \{x: \phi_2(x) \leq 1 - \varepsilon_0\}$$

satisfies

(4) $$\nu(C_0 \cap H^+(u_0, \alpha(u_0))) > 0 .$$

Assume $u_0 \in U$. Then there is a sequence $\{\rho_i\}$ with $\rho_i \to \infty$ such that $\{\rho_i u_0: i=1,\dots\} \subset \Theta_1$. Theorem 7.7 yields

(5) $$\frac{1 - \pi_{\phi_1}(\rho_i u_0)}{1 - \pi_{\phi_2}(\rho_i u_0)} \leq \frac{\int_{\bar{H}^-(u_0,\alpha(u_0))} e^{\theta\cdot x} \nu(dx)}{\varepsilon_0 \int_{C_0 \cap H^+(u_0,\alpha(u_0))} e^{\theta\cdot x}\nu(dx)}$$

$$\leq C_0 \exp(-\rho_i \varepsilon) \to 0 \quad \text{as} \quad i \to \infty .$$

Hence $\pi_{\phi_1}(\rho_i u_0) > \pi_{\phi_2}(\rho_i u_0)$ for i sufficiently large, which shows that ϕ_2 is not better than ϕ_1.

Now assume $u_0 \notin U$ but $\alpha_S(u)$ is continuous at $u_0 \in U^* - U$. Then $\varepsilon_0 > 0$ in (3) can be chosen small enough so that

(6) $$\nu(C_0 \cap H^+(u, \alpha(u))) > \varepsilon_0$$

for all $||u||=1$ with $||u-u_0|| < \varepsilon_0$. Theorem 7.7, including formula 7.7(4) for the constant c appearing in 7.7(3), now yields, for $\theta = \rho u \in N$,

$$(7)\qquad \frac{1-\pi_{\phi_1}(\rho u)}{1-\pi_{\phi_2}(\rho u)} \le \frac{\int_{\bar{H}^-(u,\alpha(u))} e^{\rho u\cdot x}\,\nu(dx)}{\int_{C_0\cap H^+(u,\alpha(u))} e^{\rho u\cdot x}\,\nu(dx)}$$

$$\le (1/\varepsilon_0)e^{-\rho\varepsilon_0}$$

for $||u|| = 1$ with $||u-u_0|| < \varepsilon_0$. $u_0 \in U^*(\Theta_1)$ implies there exists a sequence $\theta_i \in \Theta_1$ with $||\theta_i|| \to \infty$ such that $\theta_i/(||\theta_i||) \to u_0$. It follows from (7) that $\pi_{\phi_1}(\theta_i) > \pi_{\phi_2}(\theta_i)$ for i sufficiently large. Consequently ϕ_2 is not better than ϕ_1.

It follows from the two cases treated above that ϕ_2 better than ϕ_1 implies $\phi_2(x) = 1$ for (a.e.) $x \notin S$. ||

Lemma 7.13 leads directly to a criterion which can often be used to prove admissibility of conventional tests for appropriate testing problems.

7.14 Corollary

Let $\{p_\theta: \theta \in \Theta\}$, $\Theta = \Theta_0 \cup \Theta_1$ be a standard exponential family, as in 7.12. Let $\theta_2 \in R^k$ and

$$(1)\qquad S = \bigcap_{u\in U^*} \bar{H}^-(u, \alpha_S(u))$$

where $U^* = U^*(\Theta_1, \theta_2)$, as in 7.13(1). Assume (also as in 7.13) that 7.13(2) is satisfied or that $\alpha_S(u)$ is continuous at u for all $u \in U^* - U$. Let $\phi(x) = 1 - \chi_S(x)$ (= 0 if $x \in S$, =1 if $x \notin S$). Then ϕ is an admissible test.

Proof. Suppose ϕ' is any test as good as ϕ. Then, $\phi'(x) = \phi(x) = 1$ for a.e.(ν) $x \notin S$ by Lemma 7.13. But then, $\pi_{\phi'}(\theta_0) \le \pi_\phi(\theta_0)$ implies $\phi'(x) = \phi(x) = 0$ for a.e.(ν) $x \in S$. Thus, $\phi' = \phi$ a.e.(ν). It follows that ϕ is admissible. ||

Remark. It follows from Corollary 7.14 that if Θ_0 is a bounded null hypothesis and $\Theta = R^k$ then any nonrandomized test with convex acceptance region is admissible. When $\Theta_0 = \{\theta_0\}$ is simple and ν is dominated by Lebesgue measure such tests in fact form a minimal complete class -- i.e. a test is admissible if and only if it is nonrandomized and has convex acceptance region (a.e.(ν)). This is the fundamental result which was proved by Birnbaum (1955). See Exercise 7.14.3.

7.15 Application (Univariate general linear model)

Here is a customary canonical form for the normal theory general linear model: $Y \in R^p$ has the normal $N(\mu, \sigma^2 I)$ distribution, $\mu_{s+1} = \ldots = \mu_p = 0$, $\sigma^2 > 0$, and the null hypothesis to be tested is that $\mu_1 = \ldots = \mu_r = 0$, $1 \le r \le s \le p$. (See, e.g. Lehmann (1959, Chapter 7).) This can be reduced via sufficiency and change of variables to a testing question of the form considered above. Let $X_i = Y_i$, $i=1,\ldots,s$, $X_{s+1} = \sum_{j=1}^{p} Y_j^2$. Then the distributions of $X = (X_1,\ldots,X_{s+1})$ form a minimal standard exponential family with canonical parameters $\theta_i = \mu_i/\sigma^2$, $i=1,\ldots,s$, $\theta_{s+1} = -1/2\sigma^2$. The null hypothesis is, therefore, $\Theta_0 = \{\theta \in N: \theta_i = 0,\ i=1,\ldots,r\}$, so that $\Theta_1 = \{\theta \in N: \sum_{i=1}^{r} \theta_i^2 > 0\}$, where of course $N = \{\theta \in R^{s+1}: \theta_{s+1} < 0\}$.

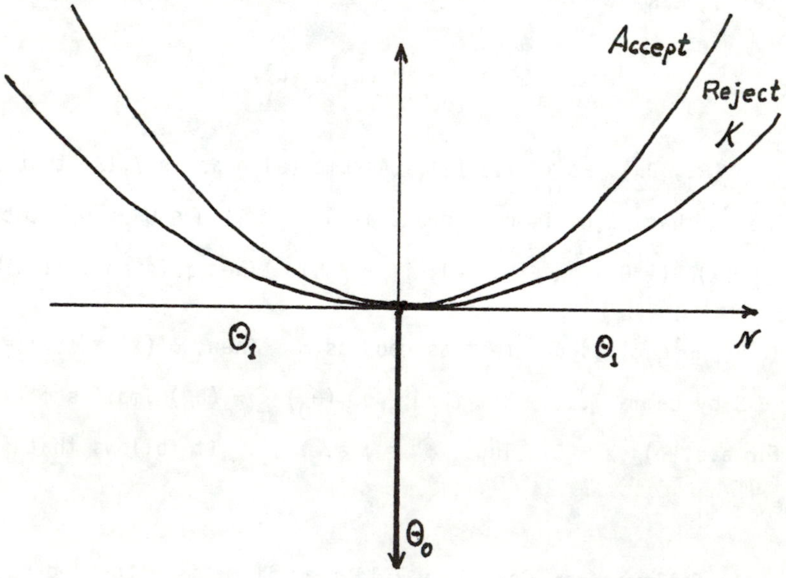

Figure 7.15(1): The F-test when $r = 1 = s$, $p = 2$.

The usual likelihood ratio F-test accepts if (and only if)

$$(1) \qquad \frac{\sum_{j=1}^{r} Y_j^2/r}{\sum_{s+1}^{p} Y_j^2/(p-s)} \le F_\alpha ,$$

as determined from tables of the F-distribution. In terms of the canonical variables this region is

$$(2) \qquad \frac{\sum_{j=1}^{r} X_j^2}{X_{s+1} - \sum_{j=1}^{r} X_j^2 - \sum_{r+1}^{s} X_j^2} \le \frac{rF_\alpha}{(p-r)} ,$$

or

$$(3) \qquad K \sum_{j=1}^{r} X_j^2 + \sum_{r+1}^{s} X_j^2 \le X_{s+1} \quad \text{where} \quad K = 1 + (p-s)/rF_\alpha > 1.$$

(The simple situation for $r = 1 = s$, $p = 2$ is illustrated in Figure 7.15(1), above, which shows K in the upper half-space and N in the lower half. Compare Figures 7.11(1) and Figure 7.12.3.)

Consider a point z in the boundary of the acceptance region (3). Thus, $K \sum_{1}^{r} z_j^2 + \sum_{r+1}^{s} z_j^2 = z_{s+1}$. The outward normal at z is $v = (2Kz_1,\ldots,2Kz_r, 2z_{r+1},\ldots,2z_s, -1)$. Except for the $(s + 1 - r)$ dimensional set having $\sum_{1}^{r} z_j^2 = 0$ all positive multiples of this vector lie in Θ_1. It follows that 7.13(1) and 7.13(2) are satisfied (for any choice of $\theta_0 \in \Theta_0$). Thus the F-test (1) (or (2)) is admissible. Note that the test remains admissible by the same reasoning if Θ_1 is restricted by $\sum_{i=1}^{r} \mu_i^2 > a\sigma^2$ since then

$$\Theta_1 = \{\theta \in N: \ \sum_{i=1}^{r} \theta_i^2 > -2\, a\, \theta_{s+1}\}.$$

The same style of reasoning can be used to prove admissibility of a wide variety of tests involving the univariate and multivariate general linear model. It was used in Stein (1956b) to prove admissibility of

Hotelling's T^2 test; Giri (1977) contains a compilation of other results provable by this method, and further references.

7.16 Discussion

If a test is shown to be admissible by virtue of Theorem 7.14 this does not, in itself, constitute a strong recommendation in favor of the test. In principle the following situation may exist: there may be another test ϕ' with $\pi_{\phi'}(\theta) \le \pi_\phi(\theta)$ for all $\theta \in \Theta_0$ and with $\pi_{\phi'}(\theta) \ge \pi_\phi(\theta)$ for "most" $\theta \in \Theta_1$. It might occur that $\pi_{\phi'}(\theta_i) > \pi_\phi(\theta_i)$ for $\theta \in \Theta_1$ except when both $\pi_{\phi'}$ and π_ϕ are very nearly one. In such a case ϕ' would dominate ϕ for all practical purposes.

Of course, a procedure whose admissibility can be proved by Theorem 7.14 may also be a desirable one. The F-test of 7.15 is a good example of this. It is admissible from several perspectives in addition to that of Theorem 7.14. The most surprising of these properties is undoubtedly the fact that it is a Bayes test. See Kiefer and Schwartz (1965) and Exercise 7.16.2.

The F-test is also locally optimal (D-optimality) in the sense that it maximizes (among level-α tests)

$$\min_{\mu \in \Theta_0} \sigma^2 \sum_{i=1}^{r} \frac{\partial^2}{\partial \mu_i^2} \pi_\phi(\mu, \sigma^2) \quad . \tag{1}$$

See Giri and Kiefer (1964) or Giri (1977) and Exercise 7.16.3. When $r = s$ the F-test, ϕ_F, is also optimal in the sense that for any constant $c > 0$ and any level-α test ϕ

$$\min \{\pi_{\phi_F}(\mu, \sigma^2): \sum_{i=1}^{r} \mu_i^2/\sigma^2 = c^2\} \tag{2}$$

$$\ge \min \{\pi_\phi(\mu, \sigma^2): \sum_{i=1}^{r} \mu_i^2/\sigma^2 = c^2\}$$

with equality only if $\phi = \phi_F$. Note that the left side of (2) is a constant. See Brown and Fox (1974b). Brown and Fox (1974a) yields the same result for $s + 1 = r$. For $r \le s + 2$ it is only known that the (minimax) inequality (2) is valid without the (admissiblity) assertion of equality only if $\phi = \phi_F$. This (minimax) assertion follows from the Hunt-Stein theorem as stated in Lehmann (1959).

The next lemma is needed for the complete class theorems which follow it. The lemma can be viewed as an elaboration of Theorem 2.17.

7.17 Lemma

Let ω_n be a sequence of (locally finite) measures concentrated on $\Theta \subset R^k$. Then there exists a subsequence $\omega_{n'}$, a closed convex set S, and a (locally finite) measure ω concentrated on $\bar{\Theta}$ such that

$$\text{(1)} \qquad \begin{aligned} \lambda_{\omega_{n'}}(b) &\to \lambda_{\omega}(b) , && b \in S^\circ \\ \lambda_{\omega_{n'}}(b) &\to \infty , && b \notin S . \end{aligned}$$

If $\omega_{n'}$, ω, and S are as in (1) and $\theta_2 \in R^k$ then

$$\text{(2)} \qquad \bar{S} = \bigcap_{u \in U^*} \bar{H}^-(u, \alpha_S(u)) ,$$

where $U^* = U^*(\Theta, \theta_2)$. (This is similar to 7.13(1).)

Proof. The first part of the lemma is a direct consequence of Theorem 2.17. To prove (2) let $T = \bigcap_{u \in U^*} \bar{H}^-(u, \alpha_S(u))$ and suppose $y \in T^\circ$. Then for every $u \in U^*$ there is an $x(u) \in S$ such that $u \cdot x(u) > u \cdot y$.

Define N(u) by

$$\text{(3)} \qquad N(u) = \{v: \; ||v|| = 1, \; v \cdot x(u) > v \cdot y\} .$$

N(u) is a relatively open subset of the unit sphere and $u \in N(u)$. Hence $\bigcup_{u \in U^*} N(u) \supset U^*$, and there is a finite subset $u_1, \ldots, u_r \subset U^*$ such that

$$\text{(4)} \qquad \underline{\underline{N}} = \bigcup_{i=1}^{r} N(u_i) \supset U^* .$$

For convenience let $x_i = x(u_i)$. Now,

$$\text{(5)} \qquad \sup \{||\theta||: \; \theta \in \Theta , \; \frac{\theta}{||\theta||} \notin \underline{\underline{N}}\} = B < \infty ;$$

otherwise there would be a sequence $v_i \notin \underline{N}$ with $v_i \to v$ ($v \notin \underline{N}$ since $\underline{N}$ is open) and a sequence $\rho_i \to \infty$ such that $\rho_i v_i \in \Theta$, $i=1,\ldots$; but then $v \in U^* \subset \underline{N}$, a contradiction. Then

$$(6) \qquad \int e^{\theta\cdot y}\, \omega_{n'}(d\theta) \le e^{B||y||}\, \omega_{n'}(\{\theta: ||\theta|| \le B\}) + \sum_{i=1}^{r} \int e^{\theta\cdot x_i}\, \omega_{n'}(d\theta)$$

$$\le e^{B||y||}\, e^{B||x_1||}\, \lambda_{\omega_{n'}}(x_1) + \sum_{i=1}^{r} \lambda_{\omega_{n'}}(x_i)$$

by (3), (4), (5) and the simple fact that

$$\omega_{n'}(\{\theta: ||\theta|| \le B\}) \le e^{B||x_1||} \int e^{\theta\cdot x_1}\, \omega_{n'}(d\theta) .$$

It follows from (6) and (1) that $y \in S$. Hence $T^\circ \subset \bar{S}$. Since T and $\bar{S}$ are closed and convex this implies $T = \bar{S}$. ||

Here is the complete class theorem from Farrell (1968). It applies to situations where Θ_0 is compact and Θ_0 and Θ_1 are separated sets. See Theorem 7.19 for a partial converse. Results like Theorem 7.18 and 7.19 have been proved in contexts somewhat more general than ordinary exponential families. See Schwartz (1967), Oosterhoff (1969), Ghia (1976), Perlman (1980), and Marden (1982a, 1982b), for such extensions and various applications. In the following statement $\bar{\Theta}_1$ denotes the closure in R^k, not merely the closure relative to N.

<u>7.18 Theorem</u>

Let $\Theta_0 \subset N$ be compact and assume $\Theta_0 \cap \bar{\Theta}_1 = \phi$. Let ϕ' be an admissible test. Then there exists an equivalent test ϕ (i.e. $\pi_{\phi'}(\theta) = \pi_\phi(\theta)$, $\theta \in \Theta_0 \cup \Theta_1$), a convex set S satisfying 7.17(2), and a (locally finite) measure H_i on $\bar{\Theta}_i$, $i=0,1$, such that $\lambda_{H_1}(x) < \infty$ for $x \in S^\circ$ and

$$\phi(x) = \begin{cases} 1 & \text{if } x \notin \bar{S} \\ 1 & \text{if } x \in S^\circ, \ \dfrac{\lambda_{H_1}(x)}{\lambda_{H_0}(x)} > 1 \\ 0 & \text{if } x \in S^\circ, \ \dfrac{\lambda_{H_1}(x)}{\lambda_{H_0}(x)} < 1, \end{cases} \tag{1}$$

a.e.(ν). ($\lambda_{H_0}(x)$ is finite since $H_0(\Theta_0) < \infty$.) If $(\Theta_0 \cup \bar{\Theta}_1)^\circ \neq \phi$ then $\phi = \phi'$; and hence all admissible tests are of the form ϕ in (1).

Proof. If ϕ' is admissible then according to Theorem 4A.10 there exists an equivalent test ϕ and a sequence of *a priori* distributions G_n (concentrated on finite subsets of Θ) whose Bayes procedures ϕ_n (say), converge to ϕ in the topology of 4A.2. By Exercise 4A.2.1 this convergence means that $\phi_n \to \phi$ weak* -- i.e.

$$\int (\phi_n(x) - \phi(x))\, g(x)\nu(dx) \to 0 \tag{2}$$

for every ν integrable function g. A consequence of (2) is that if a subsequence of $\phi_n(x)$ converges pointwise on some (measurable) subset $T \subset K$ (say $\phi_{n'}(x) \to \lambda(x)$, $x \in T$) then the limit must be ϕ (i.e., $\phi(x) = \lambda(x)$, $x \in T$, a.e.(ν)).

Let

$$H_{in}(d\theta) = e^{-\psi(\theta)}\, G_n(d\theta) \Big/ \int_{\Theta_0} e^{-\psi(\theta)}\, G_n(d\theta), \qquad \theta \in \Theta_i, \qquad i = 0, 1 \tag{3}$$

Note that $H_{0n}(\Theta_0) = 1$. Then

$$\phi_n(x) = \begin{Bmatrix} 1 \\ 0 \end{Bmatrix} \text{ if } \frac{\lambda_{H_{1n}}(x)}{\lambda_{H_{0n}}(x)} \begin{cases} > 1 \\ < 1 \end{cases}. \tag{4}$$

Let $\omega_n = H_{0n} + H_{1n}$. Let $\omega_{n'}$, ω, S be as in Lemma 7.17. Let $H_i = \omega|_{\Theta_i}$, i=0, 1, so that $H_{in'} \to H_i$, i=0, 1, as $n' \to \infty$. Then $H_0(\Theta) = H_0(\Theta_0) = 1$ since

$H_{0n}(\Theta_0) = 1$ and Θ_0 is compact. The assertions in (1) follow from this along with Lemma 7.17, (4), and the decision theoretic facts in the first paragraph of the proof.

If ϕ' and ϕ are equivalent and $(\Theta_0 \cup \bar{\Theta}_1)° \neq \phi$ then $\phi' = \phi$ (a.e.(ν)) by completeness (Theorem 2.12). ||

Many of the tests produced by the recipe 7.18(1) are admissible. In certain statistical situations, it can even be concluded that all of them are admissible. Then Theorem 7.18 describes the minimal complete class. The following converse to Theorem 7.18 contains statements of these facts. It is not entirely satisfactory but it is the best general result we have been able to devise. For the purpose of this theorem define

(1) $$\Theta_1^* = \{\theta_1 \in \bar{\Theta}_1 : \theta_1 \in N \quad \text{or there is a} \quad \theta_2 \in \Theta_1 \ni (1-\rho)\theta_2 + \rho\theta_1 \in \Theta_1 \quad \text{for} \quad 0 \leq \rho < 1\} .$$

(See Exercise 7.19.3 for an extension of (1).)

7.19 Theorem

Consider the testing problem described in Theorem 7.18.

Suppose ϕ satisfies 7.18(1) where H_1 is concentrated on Θ^* and $\bar{S}$ satisfies all the assumptions of Lemma 7.13 relative to some $\theta_2 \in R^k$. Suppose also that

(2) $$\phi(x) = \left\{\begin{matrix} 1 \\ 0 \end{matrix}\right\} \quad \text{if} \quad x \in \bar{S} \quad \text{and} \quad \frac{\lambda_{H_1}(x)}{\lambda_{H_0}(x)} \left\{\begin{matrix} > 1 \\ < 1 , \end{matrix}\right. \quad \text{a.e.}(\nu) .$$

(This is a mild extension of the latter part of 7.18(1).) Then any critical function as good as ϕ must also satisfy (2) and 7.18(1) with the same values of S, H_0, H_1. If also either

(3') $$\nu(\{x: \frac{\lambda_{H_1}(x)}{\lambda_{H_0}(x)} = 1 \quad \text{and} \quad \phi(x) < 1\}) = 0 ,$$

or

(3") $$\nu(\{x: \frac{\lambda_{H_1}(x)}{\lambda_{H_2}(x)} = 1 \text{ and } \phi(x) > 0\}) = 0 \quad ,$$

or

(4) $$(\text{Supp }(H_0 + H_1))^\circ \neq \phi$$

then ϕ is admissible; and if η is as good as ϕ then $\eta = \phi$ a.e.(ν).

If ν is dominated by Lebesgue measure, $U(\Theta, \theta_2) = U^*(\Theta, \theta_2)$ for some $\theta_2 \in R^k$, and $\bar{\Theta}_1 \subset \Theta_1^*$ then the collection of tests of the form 7.18(1) is a minimal complete class.

Proof. Suppose ϕ is defined by (2) and 7.18(1) where $\bar{S}$ satisfies the assumptions of Lemma 7.13. Suppose η is another critical function as good as ϕ. Then $\eta(x) = 1$ if $x \notin \bar{S}$ by Lemma 7.13.

If $\theta \in \Theta_1$ then

(5) $$0 \leq \int(\eta(x) - \phi(x))e^{\theta\cdot x}\,\nu(dx)$$

since $0 \leq \pi_\eta(\theta) - \pi_\phi(\theta)$. By continuity (5) also holds if $\theta \in \bar{\Theta}_1 \cap N$. Now, suppose $\zeta_\rho = (1 - \rho)\theta_2 + \rho\theta_1 \in \Theta_1$ for $0 \leq \rho < 1$. Then (5) holds at $\theta = \zeta_\rho$ and $\int(\eta(x) - \phi(x))e^{\zeta_\rho\cdot x}\,\nu(dx)$ is continuous in ρ as $\rho \uparrow 1$ by Exercise 1.13.1(ii). It follows that (5) holds whenever $\theta \in \Theta_1^*$.

The opposite inequality to (5) holds when $\theta \in \Theta_0$, and H_0 is finite since $\bar{\Theta}_0 \subset N$ is compact. Hence

(6) $$0 \leq \int (\int(\eta(x) - \phi(x))e^{\theta\cdot x}\,\nu(dx))(H_1(d\theta) - H_0(\theta)) \quad .$$

Notice that $\eta(x) - \phi(x) \leq 0$ whenever $\lambda_{H_1}(x) > \lambda_{H_0}(x)$, so that

$$\int (\eta(x) - \phi(x))^+\,\lambda_{H_1}(x)\,\nu(dx) \leq \int(\eta(x) - \phi(x))^+\,\lambda_{H_0}(x)\,\nu(dx)$$

$$= \iint e^{\theta\cdot x}\,\nu(dx)\,H_0(d\theta) < \infty \quad .$$

Furthermore, as already noted, H_0 is a finite measure. Hence the order of integration in (6) can be reversed, yielding that

$$(7) \qquad 0 \leq \int_{\bar{S}} (\eta(x) - \phi(x))(\lambda_{H_1}(x) - \lambda_{H_0}(x))\nu(dx) < \infty \quad ;$$

with the integral extending only over the region $x \in \bar{S}$ since $\eta(x) = \phi(x)$ for $x \notin \bar{S}$. Because ϕ satisfies (2), the integrand in (7) is non-positive; hence $\eta(x)$ also satisfies (2), for otherwise the integral would be negative.

If in addition ϕ satisfies (3') then $\pi_\phi(\theta_1) > \pi_\eta(\theta_1)$, $\theta_1 \in \Theta_1$, (a contradiction) unless $\eta(x) = \phi(x)$ a.e. (ν). Similarly if (3") is satisfied $\eta(x) = \phi(x)$ a.e.(ν); for otherwise $\pi_\phi(\theta_0) < \pi_\eta(\theta_0)$, $\theta_0 \in \Theta_0$. Finally, suppose (4) is satisfied in place of (3') or (3"). Note that the reasoning following (7) shows that equality holds in (7) and hence in (6). From this it follows that $\int(\eta(x) - \phi(x))e^{\theta\cdot x}\nu(dx) = 0$ a.e. $H_0 + H_1$ since this integral is non-negative on Θ_1^* and non-positive on Θ_0. (4) then implies $\eta(x) = \phi(x)$ a.e. (ν) by completeness and hence ϕ is admissible. This completes the proof of all assertions in the middle paragraph of the theorem.

If ν is dominated by Lebesgue measure and also satisfies the remaining assumptions of the last paragraph of the theorem then

$$\nu\{x: \frac{\lambda_{H_1}(x)}{\lambda_{H_0}(x)} = 1\} = 0$$

so that any test, ϕ, of the form 7.18(1) is also of the form (2), and (3') (and (3")) is satisfied, and H_1 is concentrated on $\bar{\Theta}_1 \subset \Theta_1^*$, and S satisfies assumption 7.13(2) of Lemma 7.13. It follows that ϕ is admissible. ||

COMPLETE CLASS THEOREMS FOR TESTS (Contiguous Hypotheses)

7.20 Definitions:

It is necessary to characterize the local structure of Θ_1 near Θ_0.

Let $\Theta_0 = \{\theta_0\}$ and Θ_1 be given and let

(1) $\mathcal{J}(\varepsilon) = \{J\colon J$ is a finite non-negative measure on

$$\{\theta\colon \theta \in \Theta_1, \|\theta-\theta_0\| \le \varepsilon\},\ \int J(d\theta) \le 1,\ \int \frac{J(d\theta)}{\|\theta-\theta_0\|^2} < \infty,$$

$$\left\| \int \frac{\theta-\theta_0}{\|\theta-\theta_0\|^2} J(d\theta) \right\| \le 1\}$$

Then let

(2) $$\Delta(\varepsilon) = \{(v,M)\colon v = \int \frac{\theta-\theta_0}{\|\theta-\theta_0\|^2} J(d\theta),$$

$$M = \int \frac{(\theta-\theta_0)(\theta-\theta_0)'}{\|\theta-\theta_0\|^2} J(d\theta),\ J \in \mathcal{J}(\varepsilon)\}.$$

Also, let $\Delta = \bigcap_{\varepsilon>0} \bar{\Delta}(\varepsilon)$. Note that $v \in R^k$ and M is a positive semidefinite $k \times k$ matrix, and Δ and $\bar{\Delta}(\varepsilon)$ are compact, convex sets.

In various typical statistical problems it is not hard to explicitly describe Δ. For example, if $\Theta_0 = 0$ and $\bar{\Theta} = \Theta_0 \cup \bar{\Theta}_1$ is a closed conical set then Δ is the convex hull of points of the form

(3)
$$(v,0)\colon v \in \bar{\Theta},\ \|v\| \le 1\ , \text{ and}$$
$$(0,M)\colon M = vv' \ni v \in \bar{\Theta},\ -v \in \bar{\Theta},\ \|v\| \le 1\ .$$

(See Exercise 7.20.1.) As another example, suppose Θ is a twice differentiable curved exponential family at θ_0. This means that there are two orthogonal vectors u_1, $u_2 \in R^k$, with $\|u_1\| = 1$ such that for $\theta \in \Theta$

(4) $$\theta-\theta_0 = ((\theta-\theta_0)\cdot u_1)u_1 + \|\theta-\theta_0\|^2 u_2 + o(\|\theta-\theta_0\|^2).$$

(Note in (4) that $|(\theta-\theta_0)\cdot u_1| = \|\theta-\theta_0\| + o(\|\theta-\theta_0\|^2)$, and also that $u_2 = 0$ is a possible value of u_2.) Then Δ is the convex hull of $(u_1,0)$, $(-u_1,0)$ and (u_2,u_1u_1'). (See Exercise 7.20.2.)

As with earlier results the full complete class characterization is not directly as a generalized Bayes test but involves an extension of this notion. As part of this extension the kernel $e^{\theta\cdot x}$ is replaced by

(5) $$\omega(\theta,x) = \frac{e^{\theta\cdot x}-1-\theta\cdot x}{||\theta||^2} .$$

A converse result which sometimes yields a characterization of the minimal complete class is given in Theorem 7.22. As with earlier results both of the following theorems can be profitably extended beyond the exponential family context in which they are proved below. See Marden and Perlman (1980), Marden (1981, 1982b), Cohen and Marden (1985), Brown and Sackrowitz (1984, Theorem 6.1), and Brown and Marden (in preparation).

7.21 Theorem

Let $\Theta_0 = \{\theta_0\}$ be a simple null hypothesis. Let ϕ' be an admissible test of Θ_0 versus Θ_1. Then there exists an equivalent test ϕ and a closed convex set S satisfying 7.17 (2) such that

(1) $$\phi(x) = 1 \quad x \notin S.$$

Further, for every $x_0 \in S^\circ$ there is a finite non-negative measure H on $\overline{\Theta}_1 - \{\theta_0\}$ with $S^\circ \subset \{x: e^{\theta\cdot(x-x_0)} H(d\theta) < \infty\}$, a constant $C \varepsilon R$, an $M \varepsilon \Delta_2$, and a $v \varepsilon R^k$ satisfying (3), below, with at least one of C, H, v, M being non-zero, such that for all $x \varepsilon S^\circ$

(2) $$\phi(x) = \begin{cases} 1 & \\ & \text{if } C \begin{matrix} < \\ \\ > \end{matrix} \int \omega(\theta,x-x_0)e^{\theta\cdot x_0}H(d\theta)+v\cdot(x-x_0) \\ & \qquad + (x-x_0)'M(x-x_0)/2 . \\ 0 & \end{cases}$$

If $\Theta_1^\circ \neq \phi$ then $\phi = \phi'$ a.e. (ν).

Define

$$v_\varepsilon = v - \int_{||\theta-\theta_0||>\varepsilon} \frac{\theta-\theta_0}{||\theta-\theta_0||^2} H(d\theta).$$

Then there is a sequence $\varepsilon_j \to 0$ such that $\lim_{j\to\infty} v_{\varepsilon_j} = v_0$ (say) exists, and

(3) $$(v_0,M) \varepsilon \Delta.$$

(Note that if $\int ||\theta||^{-1}H(d\theta) < \infty$ the extreme right side of (2) can be rewritten as

(3")
$$\int \frac{e^{\theta\cdot(x-x_0)}-1}{||\theta||^2} H(d\theta) + v_0\cdot(x-x_0)'M(x-x_0)/2 .$$

In particular, $\lim_{\varepsilon\to 0} v_\varepsilon = v_0$ exists.)

Proof. The assertion just after (2) follows from completeness, as in Theorem 7.17. Now, suppose ϕ' is admissible. Then by Theorem 4A.10 there is an equivalent ϕ and a sequence of prior distributions G_i concentrated on finite subsets of Θ such that the Bayes procedures, $\phi_i = \phi_{G_i}$, converge to ϕ in the topology of 4A.2. (See the proof of Theorem 7.18 for further remarks.)

Without loss of generality let $\theta_0 = 0$ and

$$G_i'(d\theta) = \frac{e^{-\psi(\theta)}G_i(d\theta)}{e^{-\psi(0)}G_i\{0\}} .$$

Thus $\phi_i(x) = \{{1 \atop 0}\}$ according to whether

(4)
$$\int_{\Theta_1} e^{\theta\cdot x}G_i'(d\theta) \gtrless 1 .$$

As in 7.17, it is possible to reduce $\{G_i'\}$ to a subsequence (if necessary) such that now for some closed S satisfying 7.17 (2),

$$\lim \int e^{\theta\cdot x}G_i'(d\theta) = \infty \qquad x \notin S$$

$$0 \le \lim \int e^{\theta\cdot x}G_i'(d\theta) = q(x) < \infty \qquad x \in S^\circ ,$$

where $G_i' \to G'$ and $q(x) = \int e^{\theta\cdot x}G'(d\theta)$. Clearly, (1) is satisfied.

Assume without loss of generality that $x_0 = 0 \in S^\circ$. Rewrite (4) as

(5)
$$\int_{\Theta_1} \omega(\theta,x)||\theta||^2G_i'(d\theta) + \int_{\Theta_1} (\theta\cdot x)G_i'(d\theta) \gtrless c_i' .$$

Let $d_i = \int||\theta||^2G_i'(d\theta) + ||\int \theta G_i'(d\theta)|| + |c_i'|$ and $H_i(d\theta) = d_i^{-1}||\theta||^2G_i'(d\theta)$. Substituting in (5) and multiplying through by d_i^{-1} yields

(6) $$\int_{\Theta_1} \omega(\theta,x)H_i(d\theta) + \int_{\Theta_1} \frac{\theta\cdot x}{\|\theta\|^2} H_i(d\theta) \gtrless c_i$$

where $\int H_i(d\theta) + \| \int(\theta/\|\theta\|^2)H_i(d\theta)\| + |c_i| = 1$. Reduce $\{H_i\}$ to a subsequence (if necessary) so that

(7) $$\int(\theta/\|\theta\|^2)H_i(d\theta) \to v, \quad c_i \to C.$$

$H_i \to H'$ since $G_i' \to G'$. Furthermore $\int H'(d\theta) + ||v|| + C = 1$ since $x_0 = 0 \in S^{\circ}$. Let $H = H'|_{\overline{\Theta}_1 - \{0\}}$.

Let $\varepsilon > 0$ such that $H(\{\theta: \|\theta\| = \varepsilon\}) = 0$. (All but a countable set of ε's satisfy this.) For each $x \in S^0$

(8) $$\int_{\|\theta\|>\varepsilon} \omega(\theta,x)H_i(d\theta) \to \int_{\|\theta\|>\varepsilon} \omega(\theta,x)H(d\theta),$$

and

(9) $$\left|\int_{\|\theta\|\le\varepsilon} (\omega(\theta,x) - \frac{x'\theta\theta'x}{2\|\theta\|^2})H_i(d\theta)\right| = O(\varepsilon)$$

since $(e^t - 1 - t - t^2/2)/t^2 = O(t)$ and $\int H_i(d\theta) \le 1$. Another subsequence may now be taken, if necessary, so that the following limits exist:

(10) $$v_\varepsilon = \lim_{i\to\infty} \int_{\|\theta\|\le\varepsilon} \frac{\theta}{\|\theta\|^2} H_i(d\theta) = v - \int_{\|\theta\|>\varepsilon} \frac{\theta}{\|\theta\|^2} H(d\theta)$$

(11) $$M_\varepsilon = \lim_{i\to\infty} \int \frac{\theta\theta'}{\|\theta\|^2} H_i(d\theta).$$

By definition, $(v_\varepsilon, M_\varepsilon) \in \overline{\Delta}(\varepsilon)$. $\overline{\Delta}(\varepsilon)$ is compact in the obvious topology. Hence there is a subsequence $\varepsilon_j \to 0$ so that $(v_{\varepsilon_j}, M_{\varepsilon_j}) \to (v_0, M) \in \Delta$. If necessary another subsequence of $\{H_i\}$ may be extracted using a diagonalization argument so that (10) and (11) hold for each ε_j. It follows from (5), (7), (8), (9), and (11) that for $x \in S^0$

$$\phi_i(x) \to \begin{matrix} 1 \\ 0 \end{matrix} \quad \text{if} \quad C \begin{matrix} < \\ > \end{matrix} \int\omega(\theta,x)H(d\theta) + v\cdot x + x'Mx/2 .$$

Note that Tr M = H'({0}). Hence Tr M + $H(\bar{\Phi}_1)$ + $||v||$ + C = 1 so that at least one of M, H, $||v||$, C are non-zero.
It follows from (10) that (3) is satisfied. Since $\phi_i \to \phi$ in the topology of 4A.2 this yields (2). ||

7.22 Theorem

Consider the testing problem described in Theorem 7.21. Suppose $\theta_0 \varepsilon N^\circ$ and ϕ satisfies 7.21(1), (2), and (3) where $\bar{S}$ satisfies all the assumptions of Lemma 7.13 and H is concentrated on Θ_1^*, as defined in 7.19(1). Suppose $\phi(x)$ is also given by 7.21(2) for $x \varepsilon \bar{S} - S^\circ$. Then any critical function as good as ϕ must also satisfy 7.21(1) for $x \notin \bar{S}$ and 7.21(2) for $x \varepsilon \bar{S}$ (a.e. (ν)) with the same values of S, H, v, M, C, x_0.

If also either

(1) $\nu\{x\colon \omega(\theta,x-x_0)H(d\theta) + v'\cdot(x-x_0) + (x-x_0)'M(x-x_0)/2 = C,\ \phi(x) < 1\} = 0$

or

(1') $\nu\{x\colon \omega(\theta,x-x_0)H(d\theta) + v'\cdot(x-x_0) + (x-x_0)'M(x-x_0)/2 = C,\ \phi(x) > 0\} = 0$

or

(2) $$(\mathrm{Supp}\ H)^\circ \neq \phi$$

then ϕ is admissible; and if η is as good as ϕ then $\eta = \phi$ a.e. (ν).

If ν is dominated by Lebesgue measure, $U(\Theta,\theta_2) = U^*(\Theta,\theta_2)$ for some $\theta_2 \varepsilon R^k$, and $\bar{\Theta}_1 = \Theta_1^*$ then the collection of tests of the form 7.21(1), (2) is a minimal complete class.

Proof. Much of the proof resembles that of Theorem 7.19 (as does much of the statement of the theorem). Assume with no loss of generality that $\theta_0 = 0$ and $x_0 = 0$. Let $\varepsilon_j \to 0$ and v_{ε_j} be as in 7.21(3) and let $J_j \varepsilon \mathcal{J}(\varepsilon_j)$, be measures supported on finite subsets such that

(3)
$$v_{\varepsilon_j} - \int \frac{\theta}{||\theta||^2} J_j(d\theta) \to 0$$
$$\int \frac{\theta\theta'}{||\theta||^2} J_j(d\theta) \to M.$$

Let $H_{1i} = H|_{\{\theta:\|\theta\|>\varepsilon_i\}} + J_i$, $H_{0i}(\{0\}) = C + \int\|\theta\|^{-2}J_i(d\theta)$. As in 7.19 if η is better than ϕ then η satisfies 7.21 (1) and

(4) $$0 \leqq \int_S (\eta(x)-\phi(x)) \int e^{\theta\cdot x}(H_{1i}(d\theta) - H_{0i}(d\theta))\nu(dx).$$

For each $x \in S$

(5) $$\begin{aligned}&\int e^{\theta\cdot x}(H_{1i}(d\theta) - H_{0i}(d\theta))\\ &= \int_{\|\theta\|>\varepsilon_i} \omega(\theta\cdot x)H(d\theta) + v\cdot x + x'Mx/2 - C\\ &\quad + \left(\int \frac{\theta}{\|\theta\|^2} J_i(d\theta) - v_{\varepsilon_i}\right)\cdot x\\ &\quad + \left(\int\omega(\theta\cdot x)J_i(d\theta) - x'Mx/2\right).\end{aligned}$$

Lemma 2.1 implies that the dominated convergence theorem can be invoked in (4), (5) as $i \to \infty$ since $0 \varepsilon N^{\circ}$ and $\omega(\theta\cdot x) = O(e^{\theta\cdot x}+1)$. Hence

(6) $$0 \leqq \int_S (\eta(x)-\phi(x))\left(\int\omega(\theta\cdot x)H(d\theta)+v\cdot x+x'Mx/2-C\right)\nu(dx).$$

It follows that η satisfies 7.21 (2). The remaining assertions of the theorem are proved just as the analogous assertions in Theorem 7.19. ||

EXERCISES

7.2.1

Prove proposition 7.2. [If ν is a finite measure and $\nu(\{||x|| > \alpha\}) = O(e^{-\varepsilon\alpha})$ then $E_\nu(e^{\varepsilon'||x||}) < \infty$ for all $0 < \varepsilon' < \varepsilon$.]

7.4.1

(i) Let S be a convex set with $\rho = \inf\{||x||: x \notin S\}$. Suppose for some $\varepsilon > 0$, $c < \infty$ 7.4(1) holds i.e.

(1) $$P_{\theta_0}(\{X/\alpha \notin S\}) < c \exp(-\varepsilon\rho\alpha) \quad \forall \ \alpha \in R .$$

Show that $\{\theta: ||\theta - \theta_0|| < \varepsilon'\}$ $N°$ for all $\varepsilon' < \varepsilon$. (ii) Give an example of a nonconvex set with $\nu\{x: ||x|| < \rho, \ x \notin S\} > 0$ and in which (1) holds but $\{\theta: ||\theta - \theta_0|| < \varepsilon'\} \not\subset N°$ for any $\varepsilon' < \varepsilon$.

7.5.1

Let $\theta_0 \in N°$ and $H^+ = H^+(v, \alpha)$. Show

(1) $$\lim_{n\to\infty} (n^{-1} \log P_{\theta_0}(\bar{X}_n \in \bar{H}^+)) \le -\tilde{K}(\bar{H}^+, \xi(\theta_0)) .$$

[Use Theorem 7.5 and Proposition 5.15.]

7.5.2

In 7.5.1 suppose $\bar{H}^+ \cap K° \neq \phi$. Show

(1) $$\lim_{n\to\infty} (n^{-1} \log P_{\theta_0}(\bar{X}_n \in H^+)) = -\tilde{K}(H^+, \ \xi(\theta_0)) .$$

[For one direction use 7.5.1(1). For the other let $P_{\tilde{\theta}}^{(n)}$ denote the distribution of S_n under $\tilde{\theta} = \theta(\tilde{\xi}_{H^+}(\theta))$.]

(2) $$P_{\theta_0}(\bar{X} \in H_0) \ge \exp[-n(K(\tilde{\theta}, \theta_0) + \varepsilon)] \, P_{\tilde{\theta}}^{(n)}(\{S: |(\theta_0 - \tilde{\theta}) \cdot \tfrac{S}{n}| < \varepsilon\})$$

$$\to \exp[-n(K(\tilde{\theta}, \theta_0) + \varepsilon)]$$

by the Central Limit Theorem (Exercise 5.15.1).]

7.5.3

Let $\theta_0 = \theta(\xi_0) \in N^\circ$. Let Q be a closed subset of R^k. Show

$$\lim_{n\to\infty} (n^{-1} \log P_{\theta_0}(\bar{X}_n \in Q)) \leq -\tilde{K}(Q, \xi_0) .$$

[Let $\varepsilon > 0$. Show $Q \subset \sum_{i=1}^{I} H^+(v_i, \alpha_i)$ where $\tilde{K}(H^+(v_i, \alpha_i)) \geq \tilde{K}(Q, \xi_0) - \varepsilon$. When $k \geq 2$ this requires some care.) Apply 7.5.2.]

7.5.4

Let $\theta_0 = \theta(\xi_0)$. Let $Q \subset R^k$ be a set such that $\tilde{K}(Q^\circ, \xi_0) = \tilde{K}(\bar{Q}, \xi_0) = \tilde{k}$ (say). Then

(1) $$\lim_{n\to\infty} n^{-1} \log P_{\theta_0}(\bar{X}_n \in Q) = -\tilde{k} .$$

[Reason as in 7.5.2 and use 7.5.3.]

7.5.5

Let $X_1, \ldots$ be i.i.d. random variables on R^k with distribution F. Let $h: X \to R^k$ be measurable and $Q \subset R^k$. Let $\xi(Q) = \inf \{\xi_F(x): x \in Q\}$ where $\xi_F(x)$ denotes the entropy as defined in 6.16(1). Suppose $\xi(Q^\circ) = \xi(\bar{Q})$ and $E(\exp(\varepsilon||X||)) < \infty$ for some $\varepsilon > 0$. Then

$$\lim_{n\to\infty} n^{-1} \log P(\bar{X}_n \in Q) = E(Q) .$$

7.5.6

(i) Show that $\tilde{K}(\cdot, \xi_0)$ is relatively continuous on $\{x: \tilde{K}(x, \xi_0) < \infty\}$ if $\nu(K - K^\circ) = 0$, if $k = 1$, or if ν is concentrated on a countable number of points satisfying Assumptions in Theorem 6.23. If so, then for Q an open set $\tilde{K}(Q, \xi_0) = \tilde{K}(\bar{Q}, \xi_0)$ as required in 7.5.4. (ii) Given an example where Q is open and $\tilde{K}(Q, \xi_0) \neq \tilde{K}(\bar{Q}, \xi_0)$. [Let ν be Lebesgue measure on the first quadrant of R^2 plus a unit mass at the origin.]

7.7.1

Hwang (1983) raises the following question: Let $X \sim N(\theta, I)$, $\theta \in R^k$. Does there exist an estimator $\delta: R^k \to R^k$ for which

$$(1) \qquad P_\theta(||\delta(X) - \theta|| \le B) \ \ge \ P_\theta(||X - \theta|| \le B) \qquad \forall\ B > 0, \quad \theta \in R^k \ ,$$

with strict inequality for some B, θ? (If so, δ would be said to "stochastically dominate" $\delta_0(x) = x$. Note that for fixed $B > 0$ there exists an estimator δ dominating δ_0 in the sense of satisfying (1) for all $\theta \in R^k$. See Hwang (*op. cit.*) and references cited therein.) It can be shown that $\delta \ne \delta_0$ exists satisfying (1) if and only if there exists a continuous spherically symmetric function $\delta \ne \delta_0$ satisfying (1). Show that no such function exists. [Suppose $||\delta(x_0)|| < ||x_0||$ for some $x_0 \in R^k$ (and hence for a neighborhood of x_0). Let $\theta_\rho = \rho x_0$ and $B_\rho = (\rho - 1)||x_0||$. Show that for some $\varepsilon > 0$, sufficiently small,

$$(1) \qquad \frac{P_{\theta_\rho}(||\delta(X) - \theta_\rho|| > B_\rho)}{P_{\theta_\rho}(||X - \theta_\rho|| > B_\rho)}$$

$$\ge \frac{P_{\theta_\rho}(X \in H^+(x_0, ||x_0||^2),\ \delta(X) \in H^-(x_0, ||x_0||^2))}{e^{\varepsilon\rho}\, P_{\theta_\rho}(X \in H^-(x_0, ||x_0||^2))} \to \infty \ .$$

Use the multivariate generalization of 7.3(3) to estimate the denominator on the left of (1); then use 7.7(3) for the asymptotic assertion in (1). A similar argument, with different θ_ρ and B_ρ, applies when $||\delta(x_0)|| > ||x_0||$ for some $x_0 \in R^k$. See Brown and Hwang (in preparation).]

<u>7.9.1</u>

Consider the estimation problem described in Exercise 4.24.3. Show that the estimtor 4.24.3(1) is admissible. [Use Theorem 7.7 and Corollary 7.9 to show that if δ' is better than δ then $\delta'(x) = 0$, $x < 1$, and $\delta'(x) \le \frac{1}{2}$, $x = 1$, and symmetrically for $x \ge 2$. Among all such estimators δ minimizes the risk at $\theta = 0$.]

<u>7.9.2 (A uniform version of Corollary 7.9)</u>

Let $V_1 \subset V_2$ be subsets of the unit sphere in R^k with V_1 closed and V_2 relatively open in the unit sphere. Let

(1) $$\alpha(v) = \sup\{\alpha: K \cap H^+(v, \alpha) \neq \phi\} \quad .$$

Assume $\alpha(v) < \infty \ \forall \ v \in V_2$. Then

(i) $\alpha(v)$ is continuous for $v \in V_2$

(ii) $\forall \ \varepsilon > 0 \ \exists \ \delta > 0 \ \ni \ \nu(H^+(v, \alpha(v) - \varepsilon)) > \delta$, $v \in V_1$

(iii) $\forall \ \varepsilon > 0 \ \exists \ r_0 \ \ni$

(2) $$v \cdot \xi(rv) > \alpha(v) - \varepsilon \ \forall \ v \in V_1, \quad r > r_0 \quad .$$

7.9.3

Consider a steep exponential family. Let $K \subset \{x: x_1 \leq 0\}$, $0 \in K$, and let K be strictly convex. Let $y \in \partial K$, $y \neq 0$. Let $\theta_i \in N°$, i=1,..., such that $\xi(\theta_i) \to y$. Then, (i) $\exists \ I < \infty$, $\varepsilon > 0$, $\delta > 0$ such that $\nu(H^+(\frac{\theta_i}{||\theta_i||}, \varepsilon)) > \delta$ for all $i > I$. Hence, (ii) $\psi(\theta_i) \geq \varepsilon||\theta|| + \ln \delta$ for all $i > I$, and (iii) $\lim_{i\to\infty} \psi(\theta_i) = \infty$.

[There exist V_1, V_2 as in 7.9.2 and $\varepsilon > 0$, $\delta > 0$, satisfying $\alpha(v) < \varepsilon$, $v \in V_2$; $v \cdot y < -2\varepsilon$, $v \in V_1$; and

(1) $$\nu(H^+(v, \varepsilon)) > \delta \ \forall \ v \notin V_1 \quad .$$

(Draw pictures in R^2 to help see why the above is true. The strict convexity is important here.) Now, $||\theta_i|| \to \infty$. (Why?) Hence, $\frac{\theta_i}{||\theta_i||} \notin V_1$ for i sufficiently large, by 7.9.2(2).]

7.9.4

Consider a steep exponential family. Let $\Theta \subset N$ be relatively closed in N and assume K is strictly convex. Suppose $x \in \partial K$ but $x \notin (\xi(\Theta \cap N°))^-$. Show that $\hat{\theta}(x) \neq \phi$. (This result complements Theorem 5.7. I believe it should be possible to prove it by showing the above hypotheses imply that 5.7(1) is satisfied. However, the hint below indicates a different argument.

[Assume $x = 0 \in K \subset \{x: x_1 \leq 0\}$ (w.l.o.g.). Apply 7.9.3 to show $\lim_{||\theta||\to\infty, \theta\in\Theta} \psi(\theta) = \infty$. Now proceed as in the proof of Theorem 5.7, following

5.7(2).]

7.9.5

Consider a standard exponential family with natural parameter space N. Let $v \in R^k$ and $\alpha_0 = \sup\{\alpha: \nu(H^+(v, \alpha)) > 0\}$. Let $\theta_i = \rho_i v + \eta_i$ as in Corollary 7.9. Then

(1) $$\lim_{i\to\infty} v \cdot \nabla\psi(\theta_i) = \alpha_0 \quad .$$

Hence, there exist a $c > -\infty$ such that

(2) $$\psi(\theta_i) \geq -c + \alpha\rho_i \quad ,$$

and, consequently,

(3) $$p_{\theta_i}(x) \to 0 \quad \forall \ x \in H^-(v, \alpha_0) \quad .$$

[The key assertion, (1), is a uniform version of Theorem 3.9, since for $\eta_i \equiv \eta$ it follows immediately from that theorem. However, it seems easier to prove (1) as a consequence of Corollary 7.9. (Alternatively, one may also derive the above, as well as 7.9, through an application of convex duality, since $K^\circ = R$, etc.)]

7.11.1

In the situation in Corollary 7.11 let $\rho(\theta_i) = P_{\theta_i}(S_2)/(P_{\theta_i}(S_1))$. Construct examples (i) in which $\rho(\theta_i) \sim ||\theta_i||^{-\alpha}$, $\alpha > 0$; (ii) in which $\rho(\theta_i) \to 0$ but $||\theta_i||^{\alpha}\, \rho(\theta_i) \to \infty$ for all $\alpha > 0$; and (iii) in which $\rho(\theta_i) = O(||\theta_i||^{-\alpha})$ for all $\alpha > 0$ but $e^{-\alpha||\theta_i||}\rho(\theta_i) \to \infty$ for all $\alpha > 0$. [(i) Let $k = 1$, $\nu(\{0\}) = 1$ and $\nu(dx) = x^{\alpha-1}\,dx$ on $x > 0$.]

7.12.1

Consider a testing problem, as in 7.12 with $\Theta_0 = H(v, \alpha) \cap N$, $\Theta_1 = N - \Theta_0$, and $\Theta_0 \cap N^\circ \neq \phi$. For $z \in R^k$, let $z = z^{(1)} + z^{(2)}$ where $z^{(1)} \in H(v, \alpha)$, $z^{(2)} = \rho v \perp H(v, \alpha)$. Assume (w.l.o.g.) $\nu(R^k) = 1$. Show

(i) If ϕ' is better than ϕ then

$$(1) \qquad \int_{x^{(1)}=y} \phi(x)\nu(dx|x^{(1)} = y) = \int_{x^{(1)}=y} \phi'(x)\nu(dx|x^{(1)} = y)$$

$$y \in H(\nu, \alpha) \quad \text{a.e.}(\nu)$$

and

$$(2) \qquad \int x^{(2)}\phi(x)\nu(dx|x^{(1)} = y) = \int x^{(2)} \phi'(x)\nu(dx|x^{(1)} = y)$$

$$y \in H(\nu, \alpha), \quad \text{a.e.}(\nu) \quad .$$

(ii) Show that ϕ is admissible if and only if for some measurable functions C_i, γ_i, $i=1,2$,

$$(3) \qquad \phi(x) = \begin{cases} 1 & \text{if} \quad x^{(1)} > C_2(x^{(2)}) \\ \gamma_2(x^{(2)}) & \text{if} \quad x^{(1)} = C_2(x^{(2)}) \\ 0 & \text{if} \quad C_1(x^{(2)}) < x^{(1)} < C_2(x^{(2)}) \\ \gamma_1(x^{(2)}) & \text{if} \quad x^{(1)} = C_1(x^{(2)}) \\ 1 & \text{if} \quad x^{(1)} < C_1(x^{(2)}) \end{cases} .$$

[This is a continuation of 2.12.1 and 2.21.2.] (Matthes and Truax (1967).)

7.12.2

Prove that if ϕ is an admissible test and $Q \subset X$ with $\nu(Q) > 0$ then ϕ must also be admissible for the same problem with dominating measure $\nu_{|Q}$.

7.12.3

Let $X_1 = \bar{X}$ and $X_2 = S^2 + \bar{X}^2$ be the canonical statistics for the two-parameter exponential family generated by a $N(\mu, \sigma^2)$ random sample. (See Example 1.2.) Consider Figure 7.12.3. Draw the broken line parallel to $\mu_0 x_1 - x_2/2 = 0$ such that $\nu(R) = \nu(S)$. (ν is defined in Example 1.2.)

(i) Show that this is possible. (ii) Let ϕ_1 be the critical function for the test with acceptance region $Q' + R - S$, and let ϕ_0 be the critical function for the usual one-sided t-test, which has acceptance region $Q' = \{x_1 < 0$ or

$x_2 \geq cx_1^2\}$. Show

$$(1)\qquad \begin{aligned} E_{(\mu,\sigma^2)}(\phi_1) &< E_{(\mu,\sigma^2)}(\phi_0) & \mu &\leq 0 \\ E_{(\mu,\sigma^2)}(\phi_1) &> E_{(\mu,\sigma^2)}(\phi_0) & \mu &\geq \mu_0 \end{aligned}$$

Hence ϕ_1 is a better test than ϕ_0 of

$$(2)\qquad H_0:\ \mu \leq 0 \quad \text{versus} \quad H_1:\ \mu \geq \mu_0 .$$

[$E(\phi_1 - \phi_0) = E(\chi_S - \chi_R)$. Now use Corollary 2.23.] (See Brown and Sackrowitz (1984). See also Exercise 7.14.6.)

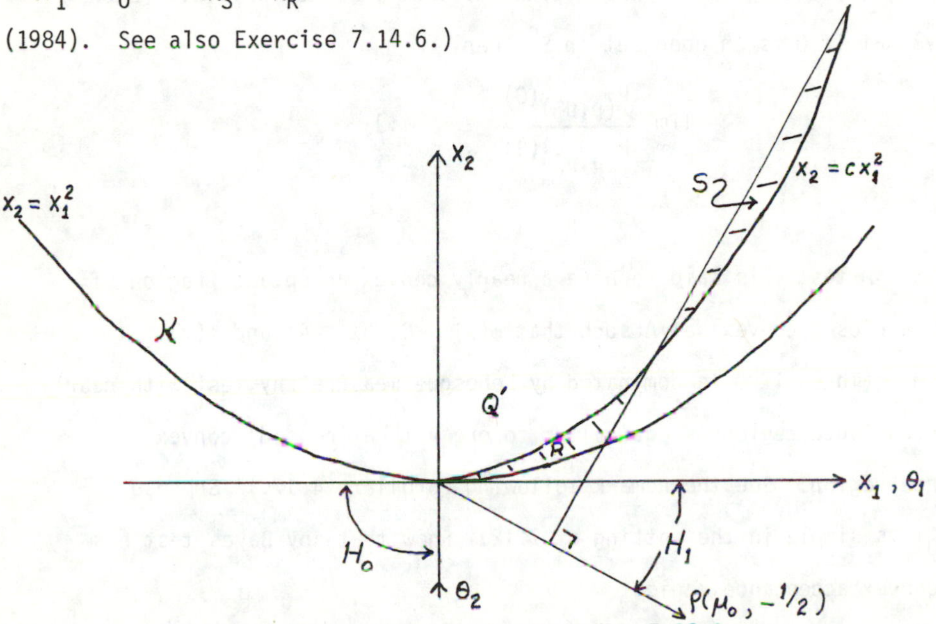

Figure 7.12.3: Diagram for Exercise 7.12.3

7.13.1

Here is an example which shows that something more than 7.13(1) is needed for validity of the conclusion of Lemma 7.13. Let $X \in R^2$ be bivariate $N(\theta, I)$. Consider the problem of testing $\Theta_0 = \{0\}$ versus $\Theta_1 = \{\theta: \theta_1 > 0, \theta_2 = -\theta_1^2\}$. Let $S = \{x \in R^2: x_2 \geq 0\}$.

(i) Show that $U = \phi$ but $U^* = (0, -1)$.

(ii) Verify that S satisfies 7.13(1) but not the remaining hypotheses of Lemma 7.13.

(iii) Let $\phi_1(x) = 1$ if $x \notin S$, $= 0$ otherwise. Show the conclusion of Lemma 7.13 does not apply to ϕ_1. [Let $\phi_2(x) = 1$ if $x_1 \geq 0$, $x_2 < \varepsilon$ or

$x_1 < 0$, $x_2 < -\varepsilon$. Show for $\varepsilon > 0$ sufficiently small ϕ_2 dominates ϕ_1.]

7.13.2

The additional assumptions of 7.13 are stronger than necessary. Let $X \sim N(\theta, I)$, $\Theta_0 = \{0\}$, S be as in 7.13.1. But now let $\Theta_1 = \{(\mu, \mu^4): \mu > 0\}$. Note that S satisfies 7.13(1) but does not satisfy either of the other two assumptions of Lemma 7.13. Show that if ϕ' is as good as ϕ then $\phi'(x) = 1$ for all $x \notin S$. Conclude that ϕ is admissible. [Show directly that if Q is an open set in S^c then

$$\lim_{\mu \to \infty} \frac{P_{(\mu,\mu^4)}(Q)}{P_{(\mu,\mu^4)}(S)} = \infty \quad .]$$

7.14.1

A test ϕ is said to have a nearly convex acceptance region if there is a closed convex set A such that $\phi(x) = 0$, $x \in A^\circ$ and $\phi(x) = 1$ for $x \notin A$. (Thus, if ν is dominated by Lebesgue measure any test with nearly convex acceptance region is equivalent to one with a (closed) convex acceptance region. See the Remark following Corollary 4.17.) Suppose $\Theta_0 = \{\theta_0\}$ is simple in the setting of 7.12. Show that any Bayes test has nearly convex acceptance region.

7.14.2

Let ϕ_i be a sequence of critical functions with nearly convex acceptance regions. Suppose $\phi_i \to \phi$ weak* on L_∞. (See 4A.2(1) for the definition of weak* convergence.) Then ϕ has a nearly convex acceptance region. [Assume $\nu(R^k) < \infty$. To each ϕ_i there corresponds an A_i. Let $\{u_j\}$ be a countable dense subset of $\{u: \ ||u|| = 1\}$. Choose a subsequence $\{i'\}$ such that $\alpha_{A_{i'}}(u_j)$ converges for each u_j, say, $\alpha_{A_{i'}}(u_j) \to \alpha_j$. Let $A = \bigcap_j \bar{H}^-(u_j, \alpha_j)$. Then $\phi(x) = 0$, $x \in A^\circ$ and $=1$ for $x \notin A$.]

7.14.3

Suppose $\Theta_0 = \{\theta_0\}$ is simple in the setting of 7.12.

(i) Show that the tests with nearly convex acceptance regions form a complete class.

(ii) Suppose, also, $\Theta_1 = R^k - \{\theta_0\}$ and ν is dominated by Lebesgue measure. Show that the tests with convex acceptance regions form a minimal complete class. [Use Theorem 4.14, 7.14.1, 7.14.2, and, for (ii), Theorem 7.14.]

<u>7.14.4</u>

Suppose the support of ν is a finite set, X. Let $\Theta_0 = \{\theta_0\} \in N = R^k$. (i) Prove that ϕ is admissible if and only if there is a closed convex set A such that $\phi(x) = 1$ if $X \notin A$, $= 0$ if $x \in A°$ or if $x \in$ r.i.F for some face F of A. (ii) Can you formulate an analogous complete class statement valid when X is countable and the assumptions of Theorem 6.23 are satisfied? [(i) Use Theorem 7.14, Corollary 7.10, and 7.12.2. (ii) Be careful; the characterization in (i) is not valid here, even when $X = \{0,1,\ldots\}^k$, and so will need to be modified.]

<u>7.14.5</u>

Consider a 2×2 contingency table. (See Exercise 1.8.1.) Two common tests for independence of row and column effects are the likelihood ratio test and the χ^2 test, based on the values of

$$\chi^2 = N \sum_{i,j} \frac{\left(Y_{ij} - \frac{Y_{i+} Y_{+j}}{N}\right)^2}{Y_{i+} Y_{+j}} .$$

(i) Use Theorem 7.14 to show that the χ^2 test is admissible.

(ii) Is the likelihood ratio test also admissible via Theorem 7.14?

(iii) Use 7.12.1 to prove both tests are admissible.

<u>7.14.6</u>

Show that the test with critical function ϕ_1 in Exercise 7.12.3 is admissible.

7.16.1

Let $X \in R^k$ be $N(\theta, I)$. Suppose $\Theta_0 = 0$ and $\Theta_1 = \{\theta: |\theta_i| > c \quad i=1,\ldots,k\}$. Consider level α tests of the form $\phi_1(x) = 1 - \chi_{\{t: |t_i| < a_1, i=1,\ldots,k\}}(x)$ and $\phi_2(x) = 1 - \chi_{\{||t|| < a_2\}}(x)$. Note that ϕ_1 is admissible. Adjust k, c, α to provide an example where ϕ_2 dominates ϕ_1 except where π_{ϕ_1} is extremely small.

7.16.2

Consider the univariate linear model, as in 7.15. Show that the usual F test, 7.15(1), is Bayes. [Let $\eta \in R^s$. Let $\sigma^2 = 1/(1 + ||\eta||^2)$ and $\mu_i = \eta_i/(1 + ||\eta||^2)$, $i = r+1,\ldots,s$. Under Θ_1 also let $\mu_i = \eta_i/(1 + ||\eta||^2)$, $i=1,\ldots,r$. Under Θ_0 (resp. Θ_1) let η have density proportional to

$$(1 + ||\eta||^2)^{-p/2} \exp\left(\frac{||\eta||^2}{2(1 + ||\eta||^2)}\right)$$

$$\left(\text{resp., } (1 + ||\eta||^2)^{-p/2} \quad \exp\left(\sum_{r+1}^{s} \frac{\eta_i^2}{2(1 + ||\eta||^2)}\right)\right).]$$

(Kiefer and Schwartz (1965).)

7.16.3

Verify when r = 2 that the F test has the local optimality property described in 7.16(1). (This is called D-optimality.) [Write

$$\Sigma \frac{\partial}{\partial \mu_i^2} \pi_\phi(0, \sigma^2) = \Sigma \int \phi(y)\left(\frac{\partial}{\partial \mu_i^2} p_\mu(y)\Big|_{\mu=0}\right)dy$$

and use a general form of the Neyman-Pearson Lemma or Theorem 2.21.]

7.16.4

Let $X_1,\ldots,X_k$ be independent gamma variables with known indices $\alpha_1,\ldots,\alpha_k$ and unknown scale parameters $\sigma_1,\ldots,\sigma_k$. Consider the problem of testing the null hypothesis H_0: $\sigma_1 = \ldots = \sigma_k$. (In the special case where the X_i/σ_i are χ^2 variables resulting from a normal sample then this is the problem of testing *homogeneity of variance*. (In this notation the variances are $\sigma_1,\ldots,\sigma_k$.)) Show

(i) The likelihood ratio test for this problem has acceptance region

$$(1) \qquad S = \{x: \frac{(\Sigma x_i)^{\alpha_0}}{\Pi x_i^{\alpha_i}} \leq C\}, \quad \text{where} \quad \alpha_0 = \sum_{i=1}^{k} \alpha_i .$$

(ii) When these distributions are written as a canonical exponential family the null hypothesis is linear in both parameter space and expectation space. Nevertheless, for $k \geq 3$, the acceptance region for the likelihood ratio test is not convex. (Hence there is no hope of proving its admissibility via Theorem 7.14.)

[(ii) Consider $k = 3$ and $\alpha_i \equiv \alpha$. Consider points of the form $x_z = (z, z, 1)$ on the boundary of the acceptance region S. Let $f(x) = \frac{\Pi x_i}{(\Sigma x_i)^3} - C$ so that $f(x) = 0$ for $x \in \partial S$. Show that for z sufficiently large $(\nabla f(x_z))' \, (D_2 f(x_z))(\nabla f(x_z)) < 0$.]

(iii) The likelihood ratio test is unique Bayes, hence admissible. Under H_1 let $\theta_i = 1/\sigma_i = (1 + \eta_i^2)$ where $\eta_i \in R$ are independent variables with density $|\eta_i|^{(\alpha_i-1)} \, (1 + \eta_i^2)^{-\alpha_i}$. Under H_0, $\theta_i = 1/\sigma_i \equiv (1 + \eta^2)$ where $\eta \in R$ has density $|\eta|^{(\alpha_0-1)}(1 + \eta^2)^{-\alpha_0}$. (This result is another one of many contained in Kiefer and Schwartz (1965).)

Note: It is not always true that a likelihood ratio test is admissible. For an interesting counter-example see Lehmann (1959, p.338) or Kiefer and Schwartz (1965, p.767).

7.17.1

Let $x \in R^2$ be bivariate normal, $N(\theta, I)$. Consider the problem of testing $\Theta_0 = \{0\}$ versus $\Theta_1 = \{\theta: \theta_1\theta_2 \geq 0 \,, \ ||\theta|| \geq 1\}$. Show that the non-randomized level $\alpha = .05$ test with acceptance region $\{x: ||x||^2 \leq 5.991\}$ is inadmissible. (Can you also find a better test?) (Compare this result with 7.22.2 in which this test is admissible.)

7.17.2

Exercise 2.10.1 indicates a nontrivial testing problem where Θ_0 and Θ_1 are contiguous and all tests are admissible. Here is an example of the same phenomenon in which the null and alternative hypotheses are separated: Let $1 \leq m < k$ and let $X = \{x \in R^k\colon x_i = 0 \text{ or } 1,\ i=1,\ldots,k,\ \Sigma x_i = m\}$. Let ν be counting measure on X, with $\{p_\theta\}$ the exponential family generated by ν. Let $\Theta_0 = \{0\}$, $\Theta_1 = \{\theta\colon ||\theta||^2 \geq 1\}$. (Other more restrictive definitions of Θ_1 will also suffice.) Let ϕ be any (possibly randomized) test. Then ϕ is admissible.

[It is possible to use Lemma 7.13 for this, but here is an easier argument. The aggregate family generated by $\{p_\theta\}$ contains $\{q_\xi\colon \xi \in X\}$ where $q_\xi(\cdot) = \chi_\xi(\cdot)$ and also $q_{\xi_0}(\cdot) \equiv \binom{k}{m}^{-1}$ where $\xi_0 = \xi(0)\ (\frac{m}{k})\underset{\sim}{1}$. If ϕ is inadmissible there exists a test ϕ' better than ϕ for testing Θ_0 versus Θ_1. Then (by continuity) ϕ' must be as good as ϕ for testing q_{ξ_0} versus $\{q_\xi\colon \xi \in X\}$. This implies $\phi'(x) \geq \phi(x)$, $x \in X$, and $\binom{k}{m}^{-1} \sum_{x \in X} \phi'(x) \leq \binom{k}{m}^{-1} \sum_{x \in X} \phi(x)$.]

7.18.1

Let X_1, X_2 be independent gamma variables $\Gamma(\alpha_i, \lambda_i)$, $i=1,2$, variables with α_1, α_2 known. Consider the problem of testing $H_0\colon \lambda_1 = \lambda_2 = 1$ versus the alternative $H_1\colon \max_{i=1,2} |1 - \lambda_i| > \varepsilon$ for some given $\varepsilon > 0$. Show that any "intersection" test with acceptance region --

(1) $\phi(x) = 0$ iff $a_{i1} < x_i < a_{i2}$, $i=1, 2$, $(0 < a_{i1} < a_{i2} < \infty)$ --

is inadmissible. (See also 7.21.1.) [No admissible test can have an acceptance region with a sharp corner at $(x_1, x_2) = (a_{12}, a_{22})$ like (1) has. See Example 2.10.]

7.19.1

In Theorem 7.19 replace Θ_1^* by

(1) $\Theta_1^{**} = \{\theta_1 \in \bar{\Theta}_1 : \theta_1 \in N$ or there is a set $\{\theta_j' : j = 1,\ldots,J\} \subset N$

and a sequence $\{\zeta_i\} \subset \Theta_1$ with $\zeta_i \to \theta_1$ and

$\{\zeta_i\} \subset \text{conhull}\,(\{\theta_j'\} \cup \{\theta_1\})\}$.

[Use 1.13.2.]

7.20.1

Prove the assertion in 7.20(3). [The extreme points of $\{J : J \in \mathcal{J}(\varepsilon), \int \theta J(d\theta) = v_0\}$, $v_0 \in \Theta$, are the distributions in this set which are concentrated on a single point; similarly the extreme points of $\{J : J \in \mathcal{J}(\varepsilon), \int \theta J(d\theta) = 0, \int \|\theta\|^2 J(d\theta) = \alpha\}$ are two-point distributions. The extreme points of $\bar{\Delta}(\varepsilon)$ are thus points (v, M) satisfying 7.19(2) with J either a one- or two-point distribution, as above. The extreme points of Δ are (contained in) the set of limits as $\varepsilon \to 0$ of these points.]

7.20.2

Prove the assertion following 7.20(4). [Let J be either a one- or two-point distribution.]

7.20.3

Generalize the assertion following 7.20(4) to apply to the situation where Θ is a twice differentiable manifold at θ_0. [First generalize 7.20(5)!]

7.21.1

In the setting of 7.18.1 consider the problem of testing $H_0 : \lambda_1 = \lambda_2 = 1$ versus the complementary alternative $H_1 : \lambda_1 \neq 1$ or $\lambda_2 \neq 1$. Show that the intersection test 7.18.1(1) is still inadmissible.

7.21.2

Consider the curved exponential family of Example 3.14 and 5.14. Let $\Theta_0 = \{\theta_0\}$ and $\Theta_1 = \Theta - \Theta_0$. To be specific take $\theta_0 = \theta(\lambda_0) = (-1,0)$; i.e., $\lambda_0 = 1$. One easily constructed test of Θ_0 is that which rejects when $|\hat{\lambda} - \lambda_0| > c_n$ with c_n chosen to give the desired level of significance. (Such a test can be constructed for any curved exponential family, and has certain asymptotic optimality properties as $n \to \infty$.) Show that for moderately large n and the usual levels of significance this test is inadmissible; although for every n there exists a (possibly very small) level of significance for which the test of this form is admissible. [Use 5.14 and Theorem 7.21. Except for small values of n or large values of c_n the acceptance region has a convex, but not strictly convex, form. Theorem 7.21 allows only very special admissible acceptance regions which are not strictly convex; and for appropriate values of n, c_n the above acceptance region is not of this special form.]

7.22.1

Let $X_1,\ldots,X_n$ be independent normal variables, $X_i \sim N(\mu, 1+\mu^2)$. Consider the problem of testing $H_0{:}\mu = 0$. Let $\phi_1 = 1$ if $|\bar{X}| > 1.96\ldots/\sqrt{n}$, $= 0$ otherwise; and $\pi_1(\mu) = E_\mu(\phi_1)$. Show

(i) ϕ_1 has level $\alpha = .05$ and is locally unbiased (i.e., $\pi_1'(0) = 0$, $\pi_1''(0) > 0$). (Is ϕ_1 also globally unbiased; i.e., $\pi_1(\mu) \geqq .05$??)

(ii) ϕ_1 is inadmissible. [Use 7.20(5) and Theorem 7.21. Note that $\theta_2 = -\frac{1}{2\sigma^2} = -(2(1+\mu^2))^{-1} \leqq -1/2$ to show ϕ_1 cannot satisfy 7.21(2) unless $H = 0$.]

(iii) Find a locally best locally unbiased level α test; i.e., the test which maximizes $\pi''(\mu)$ subject to $\pi(0) = \alpha$, $\pi'(0) = 0$. Use Theorem 7.22 to verify this test is admissible. [Admissibility actually follows directly from the fact that this test is the unique locally best locally unbiased level α test, but it may be instructive to note how this test can be written in the form 7.21(2) with $H = 0$.] Call this test ϕ_2.

((iv) Is ϕ_2 unbiased?? Is ϕ_2 better than ϕ_1?? If not, what is??)

(v) Generalize (i)-(iii) to arbitrary curved exponential families: Show that the locally unbiased test with parallel boundaries for the acceptance region is not locally best among locally unbiased tests unless $u_2 = 0$ in 7.20(5). State (convenient, frequently satisfied) conditions under which this parallel boundary test is inadmissible.

7.22.2

Let X be bivariate normal with mean θ and covariance I. Consider the problem of testing $\Theta_0 = 0$ versus $\Theta_1 = \{(\theta_1,\theta_2): \theta_1\theta_2 > 0\}$. Consider tests of the form $\phi(x) = \chi_{\{a(x_1+x_2)^2+b(x_1-x_2)^2 \geq c\}}(x)$, a,b,c > 0. (These tests are symmetric in (x_1,x_2).) Show that such a test is admissible if and only if $a \geq b$. The same result holds if $\Theta_1 = \{(\theta_1,\theta_2): \theta_1\theta_2 > 0,\ \theta_1^2+\theta_2^2 \leq 1\}$.

Appendix to Chapter 4. POINTWISE LIMITS OF BAYES PROCEDURES

This appendix contains a proof of Theorem 4.14, which was used to establish the complete class Theorems 4.16 and 4.24, and will be used again in Chapter 7. As already noted, this theorem has nothing in particular to do with exponential families, but its proof is included here since it is not readily accessible elsewhere. We will state and prove it below in a convenient form which is more general than that stated in Theorem 4.14.

4A.1 Setting

Let $\{p_\theta(x): \theta \in \Theta\}$ be any family of probability densities relative to a σ-finite measure ν on a measure space $X, \mathcal{B}$. Assume

(1) $$p_\theta(x) > 0 \qquad x \in X, \quad \theta \in \Theta \quad .$$

(This assumption is actually used only in Proposition 4A.11 and Theorem 4A.12.) Let the action space, A, be a closed convex subset of Euclidean space. The loss function is $L: \Theta \times A \to [0, \infty)$. Assume $L(\theta, \cdot)$ is a lower semicontinuous function for each $\theta \in \Theta$. Assume also that

(2) $$\lim_{||a|| \to \infty} L(\theta, a) = \infty \quad , \qquad \theta \in \Theta \; .$$

(If A is a bounded set this is trivially satisfied.) If A is bounded let $A^* = A$; if A is unbounded let $A^* = A \cup \{\underset{\sim}{i}\}$ denote the one-point compactification of A. Extend the function $L(\theta, \cdot)$ to A^* by defining

(3) $$L(\theta, \underset{\sim}{i}) = \infty \; .$$

A randomized decision procedure on A^* is a kernel $\delta(\cdot|\cdot)$ for which

$$(4)\qquad \begin{aligned}&\delta(\cdot|x) \text{ is a Borel measure on } A^*, \text{ for } x \in \mathcal{X}\\ &\delta(A|\cdot) \text{ is } \mathcal{B} \text{ measurable for each measurable set } A \subset A^* .\end{aligned}$$

A nonrandomized procedure is one for which $\delta(\cdot|x)$ is concentrated on a single point, $\delta(x)$, for almost every (ν), $x \in \mathcal{X}$. Note we use the symbol δ both for the kernel $\delta(\cdot|\cdot)$ and for the related function $\delta(\cdot)$. Let $\mathcal{D}^*$ denote the collection of all randomized decision procedures. Let $\mathcal{D} \subset \mathcal{D}^*$ denote those giving mass 1 to $A \subset A^*$, and let $\mathcal{D}_n \subset \mathcal{D}$ denote the nonrandomized procedures in $\mathcal{D}$.

As usual, the risk of any procedure is

$$(5)\qquad R(\theta, \delta) = \int_{\mathcal{X}} \int_{A^*} L(\theta, a)\, \delta(da|x)\, p_\theta(x)\nu(dx) \quad .$$

Note that $R(\theta, \delta)$ may take the value ∞. A procedure δ is admissible if

$$(6)\qquad R(\theta, \delta') \le R(\theta, \delta) \ \forall\ \theta \in \Theta \Rightarrow R(\theta, \delta') = R(\theta, \delta) \ \forall\ \theta \in \Theta \ .$$

The proof of the main result of the appendix is broken down into six main preliminary steps as follows:

(i) $\mathcal{D}^*$ is compact in an appropriate topology;

(ii) $R(\theta, \cdot)$ is lower semicontinuous on $\mathcal{D}^*$;

(iii) $\delta_i \to \delta_0$ with $\delta_i \in \mathcal{D}_n$, $i = 0,1,\ldots,$ implies $\delta_i \to \delta_0$ in measure (ν);

(iv) the minimax Theorem for finite Θ;

(v) the closure of the Bayes procedures is a complete class; and

(vi) $\mathcal{D}_n$ is a complete class when $L(\theta, \cdot)$ is strictly convex.

Formal statements of all these results and some corollaries are given below. Complete proofs are also given for all but (i) for which the reader can consult the references cited below.

4A.2 Definitions

We now define the topology on $\mathcal{D}^*$. Let $L_1 = L_1(\mathcal{X}, \mathcal{B}_{\mathcal{X}}, \nu)$ denote the Banach space of ν integrable functions. Let C^* denote the (Banach) space of continuous, real-valued functions on the compact set A^*. For every $\delta \in \mathcal{D}^*$, $f \in L_1$, $c \in C^*$ there is a number

$$\beta_\delta(f, c) = \int\int c(a)\delta(da|x)\, f(x)\nu(dx) \quad .$$

Define the topology on $\mathcal{D}^*$ according to the convergence criterion $\delta_\alpha \to \delta$ if

$$(2) \qquad \beta_{\delta_\alpha}(f, c) \to \beta_\delta(f, c) \qquad f \in L_1, \quad c \in C^* \quad .$$

(This is a "weak" topology. The collection of sets of the following form comprise a basis for this topology:

$$\{\delta \in \mathcal{D}^*: |\beta_\delta(f_i, c_j) - \beta_{\delta_0}(f_i, c_j)| < \varepsilon, \quad 1 \le i \le I, \quad 1 \le j \le J,$$
$$\delta_0 \in \mathcal{D}^*, \quad f_i \in L_1, \quad c_j \in C^*, \quad i=1,\ldots,I, \quad j=1,\ldots,J,$$
$$\varepsilon > 0\} .)$$

4A.3 Theorem

$\mathcal{D}^*$ is compact in the topology defined above.

Proof. This theorem appears in Le Cam (1955) in a form similar to the above. In a somewhat more primitive form the result appears already in Wald (1950). It is interesting to note that this theorem is actually a special case of a result in abstract functional analysis. It follows directly from Theorems V.8.6 and IV.6.3 (the Riesz representation theorem) of the classic treatise of Dunford and Schwartz (1966). For a complete, detailed proof see Farrell (1966, Appendix). ||

As has already been noted in the text, Wald's book, and the paper of Le Cam, both cited above, continue from their versions of Theorem 4A.3 and prove results similar to most of those below; but they do not explicitly state a version of Theorem 4A.12 which is our ultimate goal. See especially Wald (1950, Sections 3.5 and 3.6) and Le Cam (1955, Theorem 3.4).

4A.4 Proposition

The map $R(\theta, \cdot): \mathcal{D}^* \to [0, \infty]$ is lower semi-continuous. In other words, if $\delta_\alpha \to \delta_0$ then

(1) $$\liminf R(\theta, \delta_\alpha) \geq R(\theta, \delta_0) , \quad \theta \in \Theta .$$

Proof. Let $\delta_\alpha \to \delta_0$. Let $\theta \in \Theta$ and let $c_B(\cdot) = \min(L(\theta, \cdot), B)$. Then $c_B \in C^*$ and, for any $\delta \in \mathcal{D}^*$, $\beta_\delta(p_\theta, c_B) \uparrow R(\theta, \delta)$ as $B \uparrow \infty$. Thus,

(2) $$\lim_\alpha \inf R(\theta, \delta_\alpha) \geq \lim_\alpha \inf \beta_{\delta_\alpha}(p_\theta, c_B)$$

$$= \beta_{\delta_0}(p_\theta, c_B) \underset{B\uparrow\infty}{\uparrow} R(\theta, \delta_0) .$$

(1) follows directly from (2). ||

We will apply this proposition in roughly the following form:

4A.5 Corollary

Let $\{\theta_1, \ldots, \theta_m\} \subset \Theta$. Let $\Gamma_f \subset R^m$ be the set of available finite risk points -- i.e.

(1) $$\hat{\Gamma}_f = \{r \in R^m: \exists\, \delta \in \mathcal{D}^*, \ R(\theta_j, \delta) = r_j, \ j=1,\ldots,m\} .$$

Let $\Gamma_f \subset R^m$ be the set of points dominated by Γ_f -- i.e.

(2) $$\hat{\Gamma}_f = \{r \in R^m: \exists\, s \in \Gamma_f, \ s \leq r\}$$

where (as usual) $s \leq r$ means $s_j \leq r_j$, $j=1,\ldots,m$. Then $\hat{\Gamma}_f$ is a non-empty, convex, closed subset of R^m.

Notation: In the current context, when $R(\theta_j, \delta) < \infty$, $j=1,\ldots,m$, we write $R(\cdot, \delta)$ to denote the point $r \in R^m$ with $r_j = R(\theta_j, \delta)$, $j=1,\ldots,m$.

Proof. $R(\theta, a_0) = L(\theta, a_0) < \infty$, $\theta \in \Theta$, so $\Gamma_f \neq \phi$, and consequently $\hat{\Gamma}_f \neq \phi$. Γ_f is convex, so also $\hat{\Gamma}_f$ is convex. Suppose $r_i \in \hat{\Gamma}_f$, $i=1,\ldots,$ and $r_i \to r$. Then there exist $\delta_i \in \mathcal{D}^*$ with $R(\cdot, \delta_i) \leq r$, $i=1,\ldots$. Since $\mathcal{D}^*$

is compact there must exist a subsequence $\{i'\}$ such that $\delta_{i'}$ is convergent; $\delta_{i'} \to \delta$. Then

(3) $$(r)_j = \lim (r_{i'})_j \geq \lim\inf R(\theta_j, \delta_{i'})$$
$$\geq R(\theta_j, \delta), \quad j = 1,\ldots,m,$$

by Proposition 4A.4. It follows that $r \in \hat{\Gamma}_f$. This proves that $\hat{\Gamma}_f$ is closed. ||

Here is another useful consequence of Theorem 4A.3 and Proposition 4A.4.

4A.6 Corollary

The set of admissible procedures forms a minimal complete class.

Proof. We give a proof only for the case where $\Theta = \{\theta_1,\ldots,\theta_m\}$ is finite. The corollary will be applied in this form in the proof of Theorem 4A.10. (The proof for general Θ is basically similar, but involves some form of Zorn's lemma. See, e.g. Brown (1977).)

Let $\delta_0 \in \mathcal{D}^*$. To each $\delta \in \mathcal{D}^*$ associate the point $r(\delta) = r \in [0, \infty]^m$ for which $r_j = R(\theta_j, \delta)$, $j = 1,\ldots,m$. Let

(1) $$\bar{\Gamma} = \{r \in [0, \infty]^m: \; r = r(\delta) \;\text{ for some }\; \delta \in \mathcal{D}^*\} .$$

(This is the same as 4A.5(1), except for the fact that here $r_j = \infty$ is possible so that $r(\delta)$ is defined for all $\delta \in \mathcal{D}^*$, not merely for those δ having finite risk.) Let $\underline{r} \in \bar{\Gamma}$ be a minimal point of $\bar{\Gamma}$ which dominates $r(\delta_0)$. That is,

(2) $$\underline{r} \leq r(\delta_0); \; r \leq \underline{r} \;\text{ and }\; r \neq r \Rightarrow r \notin \bar{\Gamma} .$$

(It is shown in Lemma 4A.8 that $\underline{r}_j < \infty$, $j = 1,\ldots,m$, but that fact is not essential here.) Such a point, $\underline{r}$, can be constructed as the limit of a sequence of points $r(\delta_i) \in \Gamma$.

By Theorem 4A.3 the sequence δ_i has an accumulation point in $\mathcal{D}^*$, say $\underline{\delta}$. By Proposition 4A.4 $r(\underline{\delta}) \leq \underline{r} \leq r(\delta_0)$. Since $\underline{r}$ was minimal it follows that $\underline{\delta}$ is admissible.

It has thus been shown that any procedure, δ_0, is dominated by an admissible procedure, as asserted by the corollary. ||

4A.7 Proposition

Let $\delta_\alpha \in \mathcal{D}_n$, $\alpha = 1, \ldots,$ and suppose $\delta_\alpha \to \delta_0$ in $\mathcal{D}^*$ with $\delta_0 \in \mathcal{D}_n$. Then $\delta_\alpha(\cdot) \to \delta_0(\cdot)$ in measure (ν). Thus there is a subsequence i' for which $\delta_{i'}(\cdot) \to \delta_0(\cdot)$ a.e.(ν).

Proof. Suppose $\delta_\alpha \to \delta_0$ in $\mathcal{D}^*$ but $\delta_\alpha \not\to \delta_0$ in measure (ν). If $\delta_0 \in \mathcal{D}^*-\mathcal{D}$ then there is an $a_0 \in A$, an $\varepsilon > 0$, and a set S with $\nu(S) > 0$ such that

(1) $$|\delta_0(x) - a_0| < \varepsilon \quad \text{for all} \quad x \in S$$

and

(2) $$\limsup_\alpha \nu(\{x \in S: \ |\delta_\alpha(x) - a_0| > 2\varepsilon\}) > 0 .$$

(To verify (1) and (2) is a standard but nontrivial exercise in measure theory which uses the fact that A is a separable metric space. If $\delta_0 \in \mathcal{D}^*$ then (1) may need to be replaced by $|\delta_0(x)| > 1/\varepsilon$ and, correspondingly, the statement $|\delta_\alpha(x)| < 1/2\varepsilon$ would need to be substituted in (2). Similar substitutions would then need to be made in what follows.) Let $c \in C^*$ satisfy $0 \leq c(\cdot) \leq 1$, and

$$c(a) = \begin{cases} 1 & |a - a_0| \leq \varepsilon \\ \\ 0 & |a - a_0| \geq 2\varepsilon \end{cases} ,$$

and let $f(\cdot) = \chi_S(\cdot) \in L_1$. Then

(3) $$\beta_{\delta_0}(f, c) = \nu(S),$$

but

$$\beta_{\delta_0}(f, c) \leq \nu(\{x: \ |\delta_\alpha(x) - a_0| < 2\varepsilon\})$$

$$= \nu(S) - \nu(\{x \in S: \ |\delta(x) - a_0| \geq 2\varepsilon\})$$

so that

$$\text{(4)} \quad \liminf_\alpha \beta_{\delta_\alpha}(f,c) \leq \nu(S) - \limsup_\alpha \nu(\{x \in S: \ |\delta_\alpha(x) - a_0| \geq 2\varepsilon\})$$

$$< \ \nu(S) \ .$$

Taken together (3) and (4) contradict the assumption that $\delta_\alpha \to \delta_0$ in $\mathcal{D}^*$. This contradiction shows that $\delta_\alpha \to \delta_0$ in $\mathcal{D}^*$ implies $\delta_\alpha \to \delta_0$ in measure (ν). The second conclusion of the proposition is a standard consequence of this. ||

We now come to the minimax theorem. In preparation for this theorem we prove a simple lemma.

4A.8 Lemma

Let Θ be finite. Then the set of procedures having finite risks is a complete class of procedures in $\mathcal{D}^*$. (In other words, for every $\delta \in \mathcal{D}^*$ there is a procedure $\delta' \in \mathcal{D}^*$ with $R(\theta, \delta') \leq R(\theta, \delta)$ and $R(\theta, \delta') < \infty$, $\theta \in \Theta$.)

Proof. Let $a_0 \in A$ and $A_1 = \max\,\{L(\theta, a_0): \theta \in \Theta\} < \infty$. $B = \{a \in A: \ \min\,\{L(\theta, a): \ \theta \in \Theta\} \leq A_1\}$. B is a bounded set because of Assumption 4A.1(2). Hence

$$A_2 \ = \ \max\,\{L(\theta, a): \theta \in \Theta \ , \ \ a \in B\} \ < \ \infty \ .$$

Define δ' to satisfy

$$\text{(1)} \qquad \delta'(\{a_0\}|x) \ = \ \delta(\{a_0\}|x) + \delta(B^c|x)$$

$$\delta'(A|x) \ = \ \delta(A|x) \ , \qquad \{a_0\} \not\subset A, \qquad A \subset B \quad .$$

(In words, δ' takes action a_0 whenever δ takes an action outside B.) Then

$\delta'(B|x) \equiv 1$; hence $R(\theta, \delta') \le A_2 < \infty$. Also, by construction,

(2) $$\int L(\theta, a)\delta'(da|x) \le \int_B L(\theta, a)\delta(da|x) + L(\theta, a_0)\delta(B^c|x)$$

$$\le \int L(\theta, a)\delta(da|x) \quad .$$

Hence $R(\theta, \delta') \le R(\theta, \delta)$. ||

In the language of Corollary 4A.5, used below, the preceding can be interpreted as saying that the set of procedures with risk points in Γ_f is a complete class.

4A.9 Theorem

Let Θ be finite. Let $\delta_0 \in \mathcal{D}^*$ be any procedure for which $R(\cdot, \delta_0) \in \Gamma_f$, and such that

(1) $$R(\cdot, \delta_0) - \varepsilon \notin \hat{\Gamma}_f$$

for every $\varepsilon > 0$. Then δ_0 is Bayes -- i.e. there exists a prior G giving mass π_j to $\theta_j \in \Theta$, $j=1,\ldots,m$ such that

(2) $$\sum_{j=1}^{m} \pi_j R(\theta_j, \delta_0) \le \inf_{\delta\in\mathcal{D}^*} \sum_{j=1}^{m} \pi_j R(\theta_j, \delta) \quad .$$

Remark. The minimax risk -- $M = \inf_{\delta\in\mathcal{D}^*} \max\{R(\theta, \delta): \theta \in \Theta\}$ -- must be finite by Lemma 4A.8. (Also, as a consequence of Corollary 4A.5 there must exist a minimax procedure.) If δ_0 is any minimax procedure then it must satisfy (1) and hence must be Bayes. This does not yet prove that the resulting prior G is least favorable -- i.e. $\Sigma\, \pi_j R(\theta_j, \delta) \ge M$ for all $\delta \in \mathcal{D}^*$. Indeed, this need not be the case. To get a least favorable prior apply the proof of the theorem to the point with coordinates $r_j \equiv M$, $j=1,\ldots,m$. This point need not correspond to any procedure in $\mathcal{D}^*$, but it is in $\hat{\Gamma}$, and the proof of the theorem applies directly to yield $\{\pi_j\}$ such that $M = \sum_{j=1}^{m} \pi_j M \le \inf_{\delta\in\mathcal{D}^*} \sum_{j=1}^{m} \pi_j R(\theta_j, \delta)$. This $\{\pi_j\}$ corresponds to the least favorable distribution.

Proof. $\hat{\Gamma}_f$ is a closed convex subset of R^m by Corollary 4A.5. Condition (1) implies that the point $r_0 = R(\theta, \delta_0)$ lies on the boundary of $\hat{\Gamma}_f$. Hence there exists a nonzero vector $\{\alpha_j\}$ which defines a supporting hyperplane to $\hat{\Gamma}_f$ at r_0 -- i.e.

(3) $$\Sigma\, \alpha_j(r_0)_j = \inf\,\{\Sigma\alpha_j r_j\colon\ r \in \hat{\Gamma}_f\} .$$

Since $r_0 \in \hat{\Gamma}_f$, so also is $r_0 + ae_i$ for any unit vector e_i, and $a \geq 0$. Thus (3) yields

(4) $$\alpha_i[(r_0)_i + a] \geq \alpha_i(r_0)_i, \qquad a \geq 0 .$$

It follows that $\alpha_i \geq 0$, $i=1,\ldots,m$. Let

(5) $$\pi_i = \frac{\alpha_i}{\sum_{j=1}^{m} \alpha_j} .$$

Then

(6) $$\sum_{j=1}^{m} \pi_j(r_0)_j = \inf\,\{\Sigma\pi_j r_j\colon\ r \in \hat{\Gamma}_f\} .$$

Furthermore, by Lemma 4A.8, for every $\delta \in \mathcal{D}^*$ there is an $r \in \hat{\Gamma}_f$ such that $r_j \leq R(\theta_j, \delta)$, $j=1,\ldots,m$; so that $\sum_{j=1}^{m} \pi_j r_j \leq \Sigma\pi_j R(\theta_j, \delta)$. The desired result, (2), now follows from (6). ||

<u>4A.10 Theorem</u>

Let B_0 denote the set of Bayes procedures for priors concentrated on finite subsets of Θ. Then $\bar{B}_0$, the closure of B_0 in $\mathcal{D}^*$, is an essentially complete class.

Proof. (Note: the following proof is written in the language of directed sets, nets, and subnets. See, e.g. Dunford and Schwartz (1966). The reader unfamiliar with these concepts, or the equivalent concept of filters and ultrafilters can understand the essence of the proof by considering the case where Θ is countable, for then the nets and subnets can be converted to ordinary sequences and subsequences. If X, $\mathcal{B}$ is Euclidean space -- as in the exponential family situation -- it can be shown by an auxiliary argument that sequences and subsequences also can suffice for the proof, since the topology of $\mathcal{D}^*$ has a countable basis.) Let δ_0 be any procedure.

Let A denote the collection of all finite subsets of Θ formed into a directed set under the obvious partial ordering: $\alpha_1 \leq \alpha_2$ if $\alpha_1 \subset \alpha_2$.

Consider a fixed $\alpha \in A$; $\alpha \subset \Theta$. Consider the statistical problem with parameter space just the finite set α. There must exist a procedure, call it δ_α, which is admissible in this restricted problem and is at least as good as δ_0 -- i.e.

$$(1) \qquad R(\theta, \delta_\alpha) \leq R(\theta, \delta_0) \qquad \theta \in \alpha \quad .$$

Since δ_α is admissible in the restricted problem it satisfies condition 4A.9(1) there. (The existence of δ_α is guaranteed by Corollary 4A.6.) Hence δ_α is Bayes with respect to a prior G_α concentrated on the finite set $\alpha \subset \Theta$.

Let $A' = \{\alpha'\}$ be a (co-final) subnet of A and let $\delta \in \mathcal{D}^*$ be such that $\delta_{\alpha'} \to \delta$. (The existence of A' and δ follows from Theorem 4A.4 by standard topological arguments.) Let $\theta_0 \in \Theta$. Then $\alpha' \supset \{\theta_0\}$ for every α' far enough out in A'. Hence $R(\theta_0, \delta_{\alpha'}) \leq R(\theta_0, \delta_0)$ for any such α' and, by Proposition 4A.5,

$$R(\theta_0, \delta) \leq \liminf_{\alpha'} R(\theta_0, \delta_{\alpha'}) \leq R(\theta_0, \delta_0) \quad .$$

Since $\theta_0 \in \Theta$ is arbitrary, this proves that $\delta \in \bar{B}_0$ is as good as δ_0. Since $\delta_0 \in \mathcal{D}^*$ is also arbitrary this proves $\bar{B}_0$ is an essentially complete class. ||

So far we have not assumed that $L(\theta, \cdot)$ is strictly convex, as is the case in the applications in Chapter 4. We now add this assumption, which is required for the desired complete class theorem.

4A.11 Proposition

Assume

(1) $L(\theta, \cdot)$ is strictly convex on A for each $\theta \in \Theta$.

Let $\delta \in \mathcal{D}^*$, $\delta \notin \mathcal{D}_n$. Then there is a $\delta' \in \mathcal{D}_n$ such that

(2) $$R(\theta, \delta') \leq R(\theta, \delta), \qquad \theta \in \Theta,$$

with strict inequality for some $\theta_0 \in \Theta$. In particular, the procedures in $\mathcal{D}_n$ are a complete class.

Proof. If $\delta \in \mathcal{D}^*$ but $\delta \notin \mathcal{D}$ then $\nu(\{x\colon \delta(\{\underset{\sim}{i}\}|x)\}) > 0$. Hence $R(\theta, \delta) \equiv \infty$, $\theta \in \Theta$, by 4A.1(1). Let $a_0 \in A$ and let δ' be defined by $\delta'(x) \equiv a_0$. Then

(3) $$R(\theta, \delta') = L(\theta, a_0) < \infty = R(\theta, \delta), \qquad \theta \in \Theta.$$

Now, suppose $\delta \in \mathcal{D}$ but $\delta \notin \mathcal{D}_n$. If $R(\theta, \delta) \equiv \infty$ then, again, $\delta'(x) \equiv a_0$ satisfies (3). So, assume $R(\theta_0, \delta') < \infty$ for some $\theta_0 \in \Theta$. Condition (1) and 4A.1(2) guarantees that for some $\varepsilon > 0$, $A_0 \geq 0$

(4) $$L(\theta_0, a) \geq \varepsilon ||a|| - A_0.$$

(We leave this as an exercise on convex functions. A very similar result is proved in 5.3(3) and (5); see 5.3(4').)

Hence

(5) $$\infty > R(\theta_0, \delta) = \int(\int L(\theta, a)\delta(da|x))p_{\theta_0}(x)\,\nu(dx)$$
$$\geq \varepsilon\int(\int ||a||\,\delta(da|x))p_{\theta_0}(x)\nu(dx) - A_0.$$

It follows that

(6) $$\int ||a||\,\delta(da|x) < \infty \quad \text{a.e.}(\nu)$$

since $p_{\theta_0}(x) > 0$ a.e.(ν) by 4A.1(1).

Define

(6) $$\delta'(x) = \begin{cases} \int a\delta(da|x) & \text{if} \quad \int ||a||\delta(da|x) < \infty \\ a_0 & \text{otherwise} \end{cases} .$$

Then

$$\int L(\theta, a)\delta(da|x) \leq L(\theta, \delta'(x)) \quad \text{a.e.}(\nu)$$

with strict inequality whenever $\delta(\cdot|x)$ is not concentrated on the point $\delta'(x)$. Since $\delta \notin \mathcal{D}_n$ this occurs with positive probability under ν -- and hence by 4A.1(1) under P_{θ_0}.

Consequently

(7) $$R(\theta, \delta') \leq R(\theta, \delta)$$

with strict inequality for $\theta_0 \in \Theta$. (In fact, there is strict inequality in (7) whenever $R(\theta,\delta) < \infty$.) ||

The desired result now follows as an easy consequence.

4A.12 Theorem

Assume 4A.11(1). Then the set of pointwise limits of sequences of procedures in B_0 is a complete class. (B_0 is defined in 4A.10.)

Proof. As a consequence of 4A.11(1), Jensen's inequality and 4A.1(1) every procedure in B_0 is non-randomized. Also, there cannot be two non-equivalent admissible procedures with equal risk functions, for if $\delta_1 \neq \delta_2$ then

(1) $$(R(\theta, \delta_1) + R(\theta, \delta_2))/2 \geq R(\theta, (\delta_1 + \delta_2)/2)$$

with strict inequality whenever the right-hand side is finite.

The theorem now follows as a direct consequence of Corollary 4A.6, Proposition 4A.11, Theorem 4A.10, and Proposition 4A.7. Here's how:

Because of Corollary 4A.6 there is a unique minimal complete class. It is contained in $\mathcal{D}_n$ by Proposition 4A.11 and in B_0 by Theorem 4A.10 and (1), above. If δ_0 is in this minimal complete class (i.e., if δ_0 is admissible) there is therefore a net $\delta_\alpha \in B_0 = \mathcal{D}_n$ such that $\delta_\alpha \to \delta_0 \in \mathcal{D}_n$ in the topology on $\mathcal{D}^*$. Then, by Proposition 4A.7, $\delta_\alpha(\cdot) \to \delta_0(\cdot)$ a.e.(ν) which is the desired condition. ||

4A.13 Generalizations

(i) Assumption 4A.1(1) and the strict convexity assumption 4A.11(1) are used in the proof of Theorem 4A.12 for only two purposes; namely, to guarantee that

(1) $$\delta \in B_0 \Rightarrow \delta \in \mathcal{D}_n ,$$

and that

(2) $$(\delta_1, \delta_2 \text{ admissible}; \quad R(\theta, \delta_1) = R(\theta, \delta_2), \quad \theta \in \Theta) \Rightarrow \delta_1 = \delta_2 .$$

If (1) and (2) can be established separately, as is the case in some of the applications in Section 7 to the theory of hypotheses tests, then the conclusion of Theorem 4A.12 remains valid without 4A.1(1) and 4A.11(1).

(ii) There is not much hope for something like the conclusion of Theorem 4A.12 unless (1) and (2) are satisfied. However, all the earlier results of this appendix, through Theorem 4A.10, remain valid without the assumptions 4A.1(1) and 4A.11(1) (or (1) and (2)).

(iii) The remaining assumption which can be relaxed without major alterations in the theory is the assumption 4A.1(2) on the loss function. If this assumption is replaced by

(3) $$\lim_{a \to \underset{\sim}{\iota}} L(\theta, a) = \sup \{L(\theta, a): \ a \in A\}$$

and

(4) $$\sup \{L(\theta, a): \ a \in A\} < \infty$$

then all results through Theorem 4A.10 remain valid with only a simple modification needed in the statement and proof of Lemma 4.8 to establish that the procedures in $\mathcal{D}$ are a complete class.

(iv) If (3) is valid but not (4) or 4A.1(2), then a peculiar situation may arise. The results through Corollary 4A.7 remain valid, but it is then possible that there may exist admissible procedures having $R(\theta, \delta) = \infty$ for some $\theta \in \Theta$. When this peculiarity occurs the minimax theorem is not valid in the strong form of Theorem 4A.9. (There may exist admissible minimax procedures satisfying 4A.9(1) for which no prior exists satisfying 4A.9(2).) A weaker form of Theorem 4A.9 is, however, valid. Its conclusion is that there exists a sequence of priors defined by $\{\pi^{(\ell)}, \ell=1,\ldots\}$ and corresponding Bayes procedures $\delta^{(\ell)}$ such that $R(\theta, \delta^{(\ell)}) \to R(\theta, \delta_0)$, $\theta \in \Theta$. (The most convenient proof I know of this fact proceeds in a somewhat roundabout fashion using a device found in Wald (1950).)

(v) Brown (1977) contains versions of Theorem 4A.3 and Proposition 4A.4 valid for some situations where it is useful to compactify A in some fashion other than the one point compactification, A^*, used above; or where the loss L depends on the observed $x \in X$, as well as on Θ, A; or where the decision rules are restricted *a priori* to lie in some proper subset of $\mathcal{D}$. In many of these situations it is possible to proceed further and also establish the conclusion of Theorem 4A.10.

It is also possible to derive some satisfactory results in the (unusual) situation where A is not a Borel subset of Euclidean space, nor imbeddable as such a subset. Such an extension involves intricacies not present in the preceding treatment of the Euclidean case.

Exercises

4A.2.1

Suppose $A = \{a_0, a_1\}$, corresponding to a hypothesis testing problem. (a_0 = "accept", a_1 = "reject".) For any procedure δ let $\phi(x) = \phi_\delta(x) = \delta(\{a_1|x\})$ denote the critical function of the test. Then, $\delta_i \to \delta$ in the topology on $\mathcal{D}^*$ if and only if $\phi_{\delta_i} \to \phi_\delta$ in the weak* topology on L_∞ (i.e.

$$(1) \qquad \int |\phi_{\delta_i}(x) - \phi_\delta(x)| f(x)\nu(dx) \to 0$$

for every $f \in L_1$. (See e.g. Lehmann (1959, Section A4).)

REFERENCES

AMARI, S. (1982). Differential geometry of curved exponential families -- curvature and information loss. Ann. Statist. 10, 357-385.

ARNOLD, S.F. (1981). The Theory of Linear Models. Wiley: New York.

BAHADUR, R.R. and ZABELL, S.L. (1979). Large deviations of the sample mean in general vector spaces. Ann. Prob. 7, 587-621.

BAR-LEV, S.K. (1983). A characterization of certain statistics in exponential models whose distribution depends on a sub-vector of parameters only. Ann. Statist. 11, 746-752.

BAR-LEV, S.K., and ENIS, P. (1984). Reproducibility and natural exponential families with power variance functions. Preprint. Dept. of Statistics, S.U.N.Y., Buffalo.

BAR-LEV, S.K. and REISER, B. (1982). An exponential subfamily which admits UMPU tests based on a single test statistic. Ann. Statist. 10, 979-989.

BARNDORFF-NIELSEN, O. (1978). Information and Exponential Families in Statistical Theory. Wiley: New York.

BARNDORFF-NIELSEN, O. and BLAESILL, P. (1983a). Exponential models with affine dual foliations. Ann. Statist. 11, 753-769.

BARNDORFF-NIELSEN, O., AND BLAESILL, P. (1983b). Reproductive exponential families. Ann. Statist. 11, 770-782.

BARNDORFF-NIELSEN, O. and COX, D.R. (1984). The effect of sampling rules on likelihood statistics. Inter. Statist. Rev. 52. To appear.

BARNDORFF-NIELSEN, O. and COX, D.R. (1979). Edgeworth and saddle point approximation with statistical applications (with discussion). J. Roy. Statist. Soc. B 41, 279-312.

BASU, D. (1955). On statistics independent of a complete sufficient statistic. Sankhya 15, 377-380.

BASU, D. (1958). On statistics independent of a complete sufficient statistic. Sankhya 20, 223-226.

BERAN, R.J. (1979). Exponential models for discrete data. Ann. Statist. 7, 1162-1178.

BERGER, J.O. (1982). Selecting a minimax estimator of a multivariate normal mean. Ann. Statist. 10, 81-92.

BERGER, J.O. (1980a). Statistical Decision Theory: Foundations, Concepts, Methods. Springer-Verlag: New York.

BERGER, J.O. (1980b). Improving on inadmissible estimators in continuous exponential families with applications to simultaneous estimation of Gamma scale parameters. Ann. Statist. 8, 545-571.

BERGER, J.O. (1976). Admissible minimax estimation of a multivariate normal mean with arbitrary quadratic loss. Ann. Statist. 4, 223-226.

BERGER, J.O. (1975). Minimax estimation of location vectors for a wide class of densities. Ann. Statist. 3, 1318-1328.

BERGER, J.O. and HAFF, L.R. (1981). A class of minimax estimators of a normal mean vector for arbitrary quadratic loss and unknown covariance matrix. Mimeograph Series, Purdue University.

BERGER, J.O. and SRINIVASAN, C. (1978). Generalized Bayes estimators in multivariate problems. Ann. Statist. 6, 783-801.

BERK, R.H. (1972). Consistency and asymptotic normality of maximum likelihood estimates for exponential models. Ann. Math. Statist. 43, 193-204.

BHATTACHARYA, R.N. and RAO, R.R. (1976). Normal Approximations and Asymptotic Expansions. Wiley: New York.

BIRNBAUM, A. (1955). Characterizations of complete classes of tests of some multiparameter hypotheses with applications to likelihood ratio tests. Ann. Math. Statist. 26, 21-36.

BISHOP, Y.M., FEINBERG, S.E., and HOLLAND, P.W. (1975). Discrete Multivariate

Analysis. M.I.T. Press, Cambridge.

BLAESILD, P. (1978). A generalization of the exponential distribution to convex sets in R^k. Scand. Jour. Statist. 5, 189-194.

BROWN, L.D. (1986). Information inequalities for the Bayes risk. Statistics Center Preprint, Cornell University.

BROWN, L.D. (1981). A complete class theorem for statistical problems with finite sample spaces. Ann. Statist. 9, 1289-1300.

BROWN, L.D. (1980). Examples of Berger's phenomenon in the estimation of independent normal means. Ann. Statist. 8, 572-585.

BROWN, L.D. (1979). A heuristic method for determining admissibility of estimators -- with applications. Ann. Statist. 7, 960-994.

BROWN, L.D. (1977). Closure theorems for sequential design processes. Statistical Decision Theory and Related Topics II, Academic Press, 57-91.

BROWN, L.D. (1971). Admissible estimators, recurrent diffusions, and insoluble boundary value problems. Ann. Math. Statist. 42, 855-904.

BROWN, L.D. (1968). Inadmissibility of the usual estimators of scale parameters in problems with unknown location and scale parameters. Ann. Math. Statist. 39, 29-48.

BROWN, L.D. (1966). On the admissibility of invariant estimators of one or more location parameters. Ann. Math. Statist. 37, 1087-1136.

BROWN, L.D. (1965). Optimal policies for a sequential decision process. J. Soc. Indust. Appl. 13, 37-46.

BROWN, L.D., COHEN, A., and STRAWDERMAN, W. (1976). A complete class theorem for strict monotone likelihood ratio with applications. Ann. Statist. 4, 712-722.

BROWN, L.D. and FARRELL, R. (1984). Complete class theorems for estimation of multivariate Poisson means and related problems. Ann. Statist., to appear.

BROWN, L.D. and FOX, M. (1974a). Admissibility of procedures in two-dimensional location parameter problems. Ann. Statist. 2, 248-266.

BROWN, L.D. and FOX, M. (1974b). Admissibility in statistical problems involving a location or scale parameter. Ann. Statist. 2, 807-814.

BROWN, L.D. and HWANG, J.T. (1982). A unified admissibility proof. Stat. Dec. Theory and Related Topics, III. 1, 205-230. Academic Press: New York.

BROWN, L.D., JOHNSTONE, I.M., and MACGIBBON, K.B. (1981). Variation diminishing transformations: a direct approach to total positivity and its statistical applications. Jour. Amer. Statist. Assoc., 76, 824-832.

BROWN, L.D. and RINOTT, Y. (1985). Stochastic order relations for multivariate infinitely divisible distributions. Statistics Center Preprint, Cornell University.

BROWN, L.D. and SACKROWITZ, H. (1984). An alternative to Student's t test for problems with indifference zones. Ann. Statist. 12, 451-469.

CHERNOFF, H. (1952). A measure of asymptotic efficiency for tests of a hypothesis based on the sum of observations. Ann. Math. Statist. 23, 493-507.

CLEVINSON, M. and ZIDEK, J. (1977). Simultaneous estimation of the mean of independent Poisson laws. J. Amer. Statist. Assoc. 70, 698-705.

COURANT, R. and HILBERT, D. (1953). Methods of Mathematical Physics (Translated and revised from the German original). Interscience: New York.

COX, D.R. (1975). Partial likelihood. Biometrika, 62, 269-276.

DARROCH, J.N., LAURITZEN, S.L., and SPEED, T.P. (1980). Markov fields and log linear interaction models for contingency tables. Ann. Statist. 8, 522-539.

DARMOIS, G. (1935). Sur les lois de probabilité à estimation exhaustive. Compt. Rend. Acad. Sci., Paris 260, 1265-1266.

DE GROOT, M.H. (1970). Optimal Statistical Decisions. McGraw-Hill: New York.

DEY, D.K., GHOSH, M., and SRINIVASAN, C. (1983). Simultaneous estimation of parameters under Stein's loss. Preprint.

DIACONIS, P. and YLVISAKER, D. (1979). Conjugate priors for exponential families. Ann. Statist. 7, 269-281.

DUNFORD, N., and SCHWARTZ, J.T. (1966). Linear Operators, Part I. Interscience: New York.

DYNKIN, E.B. (1951). Necessary and sufficient conditions for a family of probability distributions. Select. Transl. Math. Statist. and Prob. 1, 23-41.

EATON, M.L. (1982). A review of selected topics in multivariate probability inequalities. Ann. Statist. 10, 11-43.

EFRON, B. (1978). The geometry of exponential families. Ann. Statist. 6, 362-376.

EFRON, B. (1975). Defining the curvature of a statistical problem (with applications to second order efficiency); with discussion. Ann. Statist. 3, 1189-1242.

EFRON, B. and TRUAX, D. (1968). Large deviations theory in exponential families. Ann. Math. Statist. 39, 1402-1424.

EQUELI, S. (1983). Second order efficiency of minimum contrast estimators in a curved exponential family. Ann. Statist. 11, 793-803.

ELLIS, R.S. (1984a). Large deviations for a general class of random vectors. Ann. Prob. 12, 1-12.

ELLIS, R.S. (1984). Entropy, Large Deviations, and Statistical Mechanics Springer-Verlag.

FARRELL, R.H. (1968). Towards a theory of generalized Bayes tests. Ann. Math. Statist. 38, 1-22.

FARRELL, R.H. (1966). Weak limits of sequences of Bayes procedures in estimation theory. Proc. Fifth Berk. Symp. Math. Statist. Prob. 1, 83-111.

FEIGIN, P.D. (1981). Conditional exponential families and a representation theorem for asymptotic inference. Ann. Statist. 9, 597-603.

FELLER, W. (1966). An Introduction to Probability Theory and Its Applications Vol. II. Wiley: New York.

FERGUSON, T.S. (1973). A Bayesian analysis of some nonparametric problems. Ann. Statist. 1, 209-230.

FLEISS, J.L. (1981). Statistical Methods for Rates and Proportions, 2nd Edition. Wiley: New York.

GHIA, G.D. (1976). Truncated generalized Bayes tests. Ph.D. thesis. Yale University.

GHOSH, M., HWANG, J.T., and TSUI, K.W. (1983). Construction of improved estimators in multiparameter estimation for discrete exponential families (with discussion). Ann. Statist. 11, 351-374.

GHOSH, M., and MEEDEN, G. (1977). Admissibility of linear estimators in the one-parameter exponential family. Ann. Statist. 5, 772-778.

GIRI, N.C. (1977). Multivariate Statistical Inference. Academic Press: New York.

GIRI, N.C. and KIEFER, J. (1964). Local and asymptotic minimax properties of multivariate tests. Ann. Math. Statist. 35, 21-35.

HABERMAN, S.J. (1979). Analysis of Qualitative Data, Volumes I, II. Academic Press: New York.

HABERMAN, S.J. (1974). The Analysis of Frequency Data. University of Chicago Press: Chicago, Illinois.

HAFF, L.R. (1982). Solutions of the Euler-Lagrange equations for certain multivariate normal estimation problems. Preprint.

HERR, D.G. (1967). Asymptotically optimal tests for multivariate normal distributions. Ann. Math. Statist. 38, 1829-1844.

HIPP, C. (1974). Sufficient statistics and exponential famlies. Ann. Statist. 2, 1283-1292.

HODGES, J.L. JR. and LEHMANN, E.L. (1951). Some applications of the Cramer-Rao inequality. Proc. 2nd Berkeley Symp. Math. Statist. Prob., 13-22.

HOEFFDING, W. (1965). Asymptotically optimal tests for multinomial distributions. Ann. Math. Statist. 36, 369-408.

HUDSON, H.M. (1978). A natural identity for exponential families with applications in multiparameter estimation. Ann. Statist. 6, 473-484.

HWANG, J.T. (1983). Universal domination and stochastic domination-estimation simultaneously under a broad class of loss functions. Cornell Statistics Center Technical Report.

HWANG, J.T. (1982). Improving upon standard estimators in discrete exponential families with applications to Poisson and negative binomial cases. Ann. Statist. 10, 857-867.

JAMES, W. and STEIN, C. (1961). Estimation with quadratic loss. Proc. Fourth Berkeley Symposium Math. Statist. and Prob. 1, 311-319.

JOAG-DEV, K., PERLMAN, M.D. and PITT, L.D. (1983). Association of normal random variables and Slepian's inequality. Ann. Prob. 11, 451-455.

JOHANSEN, S. (1979). Introduction to the Theory of Regular Exponential Families. University of Copenhagen, Institute of Mathematical Statistics, Lecture Notes.

JOHNSON, B.R. and TRUAX, D.R. (1978). Asymptotic behavior of Bayes procedures for testing simple hypotheses in multiparameter exponential families. Ann. Statist. 6, 346-361.

JOSHI, V.M. (1976). On the attainment of the Cramer-Rao lower bound. Ann. Statist. 4, 998-1002.

JOSHI, V.M. (1969). On a theorem of Karlin regarding admissible estimates for exponential populations. Ann. Math. Statist. 40, 216-223.

KALLENBERG, W.C.M. (1978). Asymptotic Optimality of Likelihood Ratio Tests in Exponential Families. Mathematical Centre Tracts, Amsterdam.

KARLIN, S. (1968). Total Positivity (Volume I). Stanford University Press, Stanford, California.

KARLIN, S. (1958). Admissibility for estimation with quadratic loss. Ann. Math. Statist. 29, 406-436.

KARLIN, S. and RINOTT, Y. (1981). Total positivity properties of absolute value multinormal variables with applications to confidence interval estimates and related probabilistic inequalities. Ann. Statist. 9, 1035-1049.

KOOPMAN, B.O. (1936). On distributions admitting a sufficient statistic. Trans. Amer. Math. Soc. 39, 399-409.

KOUROUKLIS, S. (1984). A large deviation result for the likelihood ratio statistic in exponential families. Ann. Statist., to appear.

KOZIOL, J.A. and PERLMAN, M.D. (1978). Combining independent chi-squared tests. Jour. Amer. Statist. Assoc. 73, 753-763.

KULLBACK, S. (1959). Information Theory and Statistics. Wiley: New York.

KULLBACK, S. and LEIBLER, R.A. (1951). On information and sufficiency. Ann. Math. Statist. 22, 79-86.

LAURITZEN, S.L. (1984). Extreme point models in statistics. Scand. J. Statist. 11, 65-91.

LEHAMNN, E.L. (1983). Theory of Point Estimation Wiley: New York.

LEHMANN, E.L. (1959). Testing Statistical Hypotheses. Wiley: New York.

LE CAM, L. (1955). An extension of Wald's theory of statistical decision functions. Ann. Math. Statist. 26, 69-81.

LINDSAY, B.G. (1983). The geometry of mixture likelihoods, part II: the exponential family. Ann. Statist. 11, 783-792.

LINNIK, Y.V. (1968). Statistical problems with nuisance parameters. Amer. Math. Soc. Transl. of Math. Monographs, 20.

MANDELBAUM, A. (1984). Linear estimators and measurable linear transformations on a Hilbert space. Z. Wahr. verw. Gebiet. 65, 385-397.

MARDEN, J.I. (1983). Admissibility of invariant tests in the general multivariate analysis of variance problem. Ann. Statist. 11, 1086-1099.

MARDEN, J.I. (1982a). Combining independent noncentral chi-squared or F-tests. Ann. Statist. 10, 266-277.

MARDEN, J.I. (1982b). Minimal complete classes of tests of hypotheses with multivariate one-sided alternatives. Ann. Statist. 10, 962-970.

MARDEN, J.I. (1981). Invariant tests on covariance matrices. Ann. Statist. 9, 1258-1266.

MARDEN, J.I. and PERLMAN, M.D. (1981). The minimal complete class of procedures when combining independent non-central F-tests. Proc. Third Purdue Symp. Dec. Theory and Related Topics.

MARDEN, J.I. and PERLMAN, M.D. (1980). Invariant tests for means with covariates. Ann. Statist. 8, 25-63.

MARDIA, K.V. (1975). Statistics of directional data (with discussion). J. Roy. Statist. Soc. B. 27, 343-349.

MARDIA, K.V. (1972). Statistics of Directional Data. Academic Press: London.

MARDIA, K.V. (1970). Families of Bivariate Data. Griffin: London.

MARSHALL, A.W. and OLKIN, I. (1979). Inequalities: Theory of Marjorization and Its Applications. Academic Press: New York.

MATTHES, T.K. and TRUAX, D.R. (1967). Tests of composite hypotheses for the multivariate exponential family. Ann. Math. Statist. 38, 681-697.

MEEDEN, G. (1976). A special property of linear estimates of the mean. Ann. Statist. 4, 649-650.

MEEDEN, G., GHOSH, M. and VARDEMAN, S. (1984). Some admissible nonparametric and related finite population sampling estimators. To appear in Ann. Statist.

MORRIS, C.N. (1983). Natural exponential families with quadratic variance functions: statistical theory. Ann. Statist. 11, 515-529.

MORRIS, C.N. (1982). Natural exponential families with quadratic variance functions. Ann. Statist. 10, 65-80.

MUIRHEAD, R.J. (1982). Aspects of Multivariate Statistical Theory. Wiley: New York.

NEVEU, J. (1965). The Mathematical Foundations of the Calculus of Probability. Holden-Day: San Francisco.

NEY, P. (1983). Dominating points and the asymptotics of large deviations for random walk on R^d. Ann. Prob. 11, 158-167.

NEYMAN, J. (1938). On statistics the distribution of which is independent of the parameters involved in the original probability law of the observed variables. Stat. Res. Mem. II, 58-59.

OOSTERHOFF, J. (1969). Combination of One-Sided Statistical Tests. Mathematisch Centrum, Amsterdam.

PATIL, G.P. (1963). A characterization of the exponential type distributions. Biometrika 50, 205-207.

PING, C. (1964). Minimax estimates of parameters of distributions belonging to the exponential family. Chinese Math. 5, 277-299.

PITMAN, E.J.G. (1936). Sufficient statistics and intrinsic accuracy. Proc. Camb. Phil. Soc. 32, 567-579.

RAIFFA, H. and SCHLAIFER, R. (1961). Applied Statistical Decision Theory. Harvard University: Boston.

ROCKAFELLAR, R.T. (1970). Convex Analysis. Princeton University Press: Princeton, New Jersey.

SACKS, J. (1963). Generalized Bayes solutions in estimation problems. Ann. Math. Statist. 34, 751-768.

SAW, J.G. (1977). On inequalities in constrained random variables. Comm. Statist. Theor. Meth. A6(13), 1301-1304.

SCHWARTZ, R. (1967). Admissible tests in multivariate analysis of variance. Ann. Math. Statist. 38, 698-710.

SIMON, G. (1973). Additivity of information in exponential family probability laws. J. Amer. Statist. Assoc. 68, 478-482.

SIMONS, G. (1980). Sequential estimators and the Cramer-Rao lower bound. J. Statist. Planning and Inference 4, 67-74.

SOLER, J.L. (1977). Infinite dimensional-type statistical spaces (Generalized exponential families). Recent Developments in Statistics (edited by J.R. Barra, et al.). North Holland Publishing Co.: Amsterdam.

SRINIVASAN, C. (1984). On estimation of parameters in a curved exponential family with applications to the Galton-Watson process. Technical report, University of Kentucky.

SRINIVASAN, C. (1981). Admissible generalized Bayes estimators and exterior boundary value problems. Sankhya 43, 1-25.

STEIN, C. (1973). Estimation of the mean of a multivariate normal distribution. Proc. Prague Symp. on Asymptotic Statist., 345-381.

STEIN, C. (1956a). Inadmissibility of the usual estimator for the mean of a multivariate normal distribution. Proc. Third Berkeley Symp. Math. Statist. Prob. 1, 197-206.

STEIN, C. (1956b). The admissibility of Hotellings T^2-test. Ann. Math. Statist. 27, 616-623.

SUNDBERG, R. (1974). Maximum likelihood theory for incomplete data from an exponential family. Scand. J. Statist. 1, 49-58.

VAN ZWET, W.R., and OOSTERHOFF, J. (1967). On the combination of independent test statistics. Ann. Math. Statist. 38, 659-680.

WALD, A. (1950). Statistical Decision Functions. Wiley: New York.

WIJSMAN, R.A. (1973). On the attainment of the Cramer-Rao lower bound. Ann. Statist. 1, 538-542.

WIJSMAN, R.A. (1958). Incomplete sufficient statistics and similar tests. Ann. Math. Statist. 29, 1028-1045.

WOODROOFE, M. (1978). Large deviations of likelihood ratio statistics with applications to sequential testing. Ann. Statist. 6, 72-84.

INDEX